SQUIDs, the Josephson Effects and Superconducting Electronics

The Adam Hilger Series on Measurement
Science and Technology

SQUIDs, the Josephson Effects and Superconducting Electronics

J C Gallop

National Physical Laboratory

Adam Hilger
Bristol, Philadelphia and New York

British Library Cataloguing in Publication Data

Gallop, J. C.
SQUIDS, the Josephson effects and superconducting electronics.
1. Superconductivity
I. Title
537.623

ISBN 0-7503-0051-5

Library of Congress Cataloging-in-Publication Data

Gallop, J. C. (John C.)
SQUIDS, the Josephson effects and superconducting electronics
J. C. Gallop.
p. cm.—(The Adam Hilger series on measurement science and technology)
Includes bibliographical references and index.
ISBN 0-7503-0051-5
1. Superconducting quantum interference devices.
2. Josephson effect. I. Title. II. Series.
TK7872.S8G35 1990
621.381—dc20 90–37769
CIP

Published under the Adam Hilger imprint by IOP Publishing Ltd
Techno House, Redcliffe Way, Bristol BS1 6NX, England
335 East 45th Street, New York, NY 10017-348, USA

US Editorial Office: 1411 Walnut Street,
Philadelphia, PA 19102

Typeset by MCS Ltd, Salisbury
Printed in Great Britain by J W Arrowsmith Ltd

Contents

Introduction

The remarkable property of superconductivity was first discovered eighty years ago. The topic of this book is one part of the science and technology of superconductivity which has gained prominence over the past twenty years. It has come to be called 'superconducting electronics'. This rather mundane title obscures the fascinating and fundamental principle which underpins the whole subject, namely that superconductivity is a macroscopic quantum phenomenon. The primary aim of this book is to provide an account of the development of superconducting electronics, stressing the fundamental principles on which devices and their applications are based. It is also intended that those who may wish to use superconducting devices will find some practical help and guidance.

The book is arranged as follows: first is an introduction to the basic properties of superconductivity, treated in a phenomenological way, but with emphasis on macroscopic quantum aspects of the topic. The treatment follows in outline the theory of Ginzburg and Landau, which identifies a complex order parameter with the macroscopic quantum properties of superconductivity, providing a straightforward development of theory towards the device descriptions which follow in later chapters. Both flux quantisation and the Josephson effects will be derived from some basic assumptions of the Ginzburg–Landau theory. These two phenomena are the basis for almost all the superconducting devices which will be described. Chapter 2 deals with the Josephson effects in more detail, including applications such as a quantum voltage standard, microwave and far-infrared detectors, thermometry and tunnelling spectroscopy. Next the SQUID (Superconducting QUantum Interference Device) is introduced, a detector of unparalleled sensitivity. The operating principles of the two basic versions of SQUIDs, the single- and double-junction (or RF and DC) types, are dealt with in some detail in Chapters 3 and 4, respectively. Noise processes which restrict the available sensitivity are explained, together with some quantum mechanical limits to measurement.

SQUIDs are essentially detectors of magnetic flux, but by using appropriate transducers coupled to the input of the device, any number of physical parameters may be measured. Chapter 5 is concerned with outlining some of these transducers, with details of construction and optimisation being given where appropriate or interesting. This leads on naturally to a discussion of actual applications of SQUIDs in analogue measurements. SQUIDs have shown great promise also as the basis for a family of new high-speed, low power-consumption, digital electronic components and the present state of the art is reviewed in Chapter 6.

Because of the extreme sensitivity of SQUIDs and Josephson junctions, there are considerable practical difficulties in using superconducting electronics in real-life situations. Some hints and tips are included in the next chapter as an aid for those readers who may wish to use SQUIDs, etc, in their own work. Chapter 8 deals with a number of fascinating developments in the use of superconducting devices in the areas of fundamental particles and fundamental physics, particularly gravitation studies. Future prospects in this area are considered as well as a review of what has so far been achieved.

The steadily evolving history of superconductivity has been punctuated by a number of dramatic breakthroughs, for example the formulation of the BCS theory, type II superconductivity and the Josephson effects. Undoubtedly the most dramatic happening since the original discovery of the property has been the discovery of a completely new class of superconducting compounds, mostly based on complex copper oxides with modified perovskite structure. The first compound was discovered by Müller and Bednorz in 1986 when they raised the upper temperature limit at which superconductivity was known to occur from 23 K, where it had remained for some fifteen years, to 50 K. Further rapid advances have followed, with the record at the time of writing being 125 K. The number of workers involved in superconductivity and in superconducting electronics has grown even faster than the transition temperature and many prototype SQUIDs, etc, operating in liquid nitrogen at 77 K have already been described. The final chapter attempts to put this latest revolution into perspective, ignoring the hype which inevitably accompanied the original reports, but pointing out the new possibilities which these exciting developments have brought.

I am most grateful to very many colleagues and friends whose ideas and conversations form much of the basis of this book. Particular thanks are due to Brian Petley, Bill Radcliffe and Conway Langham for their critical reading of various versions of the manuscript, and to Caroline, Ben and Tilly Gallop for their patience over the six years it has taken me to complete this book.

J C Gallop
National Physical Laboratory

1

Introduction to Superconductivity

1.1 Basic Properties

The universally known fact about superconductivity is that below some characteristic transition temperature (T_c) many metals exhibit a complete lack of electrical resistance to a flow of direct current. Such a lossless 'supercurrent' can be set up around a closed loop of superconducting wire, there being no detectable decay with time (with a half-life shown experimentally to be at least 10^6 years).

This complete loss of resistance was first observed by Onnes (1911) for the metal mercury for which $T_c = 4.19$ K, close to the normal boiling point of liquid helium. The latter had been liquefied for the first time only shortly before in the same laboratory, thus opening up a new field of *cryogenics*.

1.1.1 Materials exhibiting superconductivity

Up to the present some 28 out of the 75 metallic elements have been shown to exhibit zero resistance with transition temperatures ranging from 0.005 K for rhodium up to 9.2 K for niobium. There are a vast number of alloy and compound superconductors. At the time of writing the highest known T_c is 125 K for a perovskite structured ternary oxide of thallium, calcium, barium and copper (see Chapter 9). The current frenetic activity to raise T_c to as high a value as possible is linked to the reduced cost of refrigeration which this would incur, and hence to a far greater application of superconducting devices in the world outside of research laboratories. Many other materials also exhibit superconductivity, including a number of non-metallic compounds and some organic salts, a few of which have T_c's which only a few years ago would have been regarded as remarkable. At the time of writing these have no apparent practical potential, although they reflect clearly the increasingly widespread but complex nature of the superconducting state.

1.1.2 Critical magnetic fields

From the earliest years the remarkable properties of superconductors have led to many proposed applications of the effects. Onnes himself suggested that high-field magnets having zero electrical losses could be made using superconductors. However the loss-free current-carrying capability of all these materials is very dependent on magnetic field, as well as on temperature. Above some critical field H_c, characteristic of the material and temperature, superconductivity is destroyed. The field generated by a current flowing in the superconductor itself must also be considered, so that even in zero external magnetic field there is a maximum supercurrent which can flow. This is a function of geometry as well as temperature and material. Onnes' original hopes for high-field solenoids took many years to realise, as the critical fields of elemental superconductors are too low to be useful. Amongst these the highest occurs for the element niobium, with a value of around 0.2 T, a factor of at least ten less than can be achieved with iron-cored solenoids.

1.1.3 Large-scale and high magnetic field applications

Some 50 years after Onnes' discovery a number of practically useful alloy superconductors, mainly based on niobium, began to be developed. These alloys have critical fields as great as 40 T, sufficient to replace high-field conventional solenoids, since the extra cost of refrigeration etc is far outweighed by the cost of power consumption through ohmic losses in normal metal coils.

As well as providing a reduction of power consumption, superconducting magnets are also able to provide very spatially and temporally stable magnetic fields. Superconducting solenoids may be operated in the 'persistent mode', in which the ends of the windings are joined so that no external current source is required. The stability of high magnetic fields produced by the best persistent supercurrent solenoids has been shown to be better than 1 part in 10^{10} per hour (Petley 1989). The ability to use much higher current densities in superconducting magnets means that solenoid windings may be tailored to produce fields which are extremely spatially uniform (to better than 1 in 10^8 over a volume of $10^{-6}\,m^3$).

Apart from magnets, superconducting windings may also be used to improve the efficiency of motors and generators. A limited number of pilot applications of this kind have already been demonstrated. A further expansion of these 'electrical engineering' applications will probably not happen unless and until there is a new round of construction of even larger electrical generating sets. The subject of this book is 'superconducting electronics', dealing with devices that rely entirely on the extreme sensitivity of superconducting circuits to very low magnetic fields. Such applications do not require vast capital investment and hence may be expected to proceed to expand in a steady way.

Superconductors are not truly lossless in applied alternating electromagnetic fields. In addition to being a strong function of temperature, particularly in the region from T_c down to roughly $T_c/3$, the losses are also a rapidly increasing function of frequency of the applied oscillating field. 'Passive' AC applications of superconductivity, which rely only on low-loss properties, have so far been limited to high-Q microwave cavities. These have been used as frequency standards, narrow band filters and, in higher-power operations, as components in particle accelerators.

1.2 The Meissner Effect

Applications of superconductivity of the electrical engineering type outlined above depend wholly on the property of zero electrical resistance. In fact, superconductivity cannot be simply represented as resulting from a normal metal which happens to have infinite conductivity. The magnetic properties of superconductors are more fundamental to an understanding of the true nature of the effects. It was not until 1933 that Meissner and Ochsenfeld demonstrated the magnetic flux exclusion effect. They showed that if a superconductor was cooled through its transition temperature in an applied field less than H_c, the magnetic flux was not 'frozen into' the metal, as would be predicted by the notion of zero electrical resistance and an application of Faraday's law, but was expelled from the bulk of the material†. In conventional terms it is as if the superconductor is a perfect diamagnet. The total flux density $\boldsymbol{B}$ in a magnetic material of susceptibility χ in an applied magnetic field $\boldsymbol{H}$ is

$$\boldsymbol{B} = \mu_0 \boldsymbol{H} + \mu_0 \chi \boldsymbol{H} \tag{1.1}$$

so that in a superconductor the Meissner effect requires that $\boldsymbol{B} = 0$ and $\chi = -1$. This macroscopic perfect diamagnetism is achieved by a supercurrent flowing with no dissipation in a thin surface layer of the material,

† Application of Faraday's and Ohm's laws ($\nabla \times \boldsymbol{E} = -\partial \boldsymbol{B}/\partial t$, $\boldsymbol{j} = \sigma \boldsymbol{E}$) to the case of a superconductor for which the conductivity $\sigma = \infty$ would imply that, if infinite currents are to be avoided, $\boldsymbol{E} = 0$ and $\partial \boldsymbol{B}/\partial t = 0$ everywhere and thus as the material makes a transition from normal to superconducting state $\Delta \boldsymbol{B} = \boldsymbol{B}_n - \boldsymbol{B}_s = 0$.

taking up such a distribution that it sets up an equal and opposite field to that which is applied at all interior points in the superconductor, so screening it out completely. Figure 1.1 shows schematically how the flux density varies through a cross section of a superconductor, with the field being attenuated in a surface layer of thickness λ (known as the magnetic penetration depth). If the external field is increased to the critical value $H_c(T)$ superconductivity is destroyed, the screening currents vanish and the field penetrates the material again.

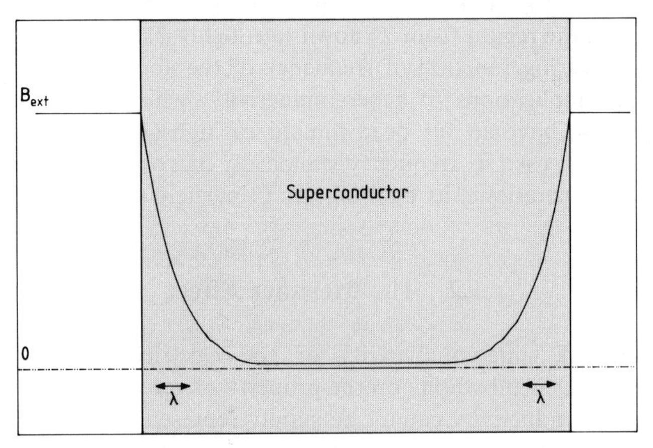

Figure 1.1 Attenuation of magnetic field into superconductor

The Meissner effect was first explained using the two-fluid model (section 1.3) and a pair of electrodynamic relationships, derived by London (1950). We shall first give another derivation based on the macroscopic quantum nature of superconductivity (section 1.6), before returning to the phenomenological description in section 1.8.

1.2.1 Temperature dependence of superconducting properties

At T_c the critical field for a superconductor is vanishingly small, rising as the temperature falls, tending asymptotically to a maximum value at $T = 0$ K. Similar behaviour is shown by the supercurrent-carrying capabilities. The penetration depth is infinite at T_c but decreases with a temperature dependence

$$\lambda(T) = \lambda(0)/[1 - (T/T_c)^4]^{1/2}. \tag{1.2}$$

These properties are interpreted simply in terms of a phenomenological theory known as the 'two-fluid model', to be outlined below in section

1.3, which simple picture explains qualitatively the temperature variation of critical currents and fields, and of the penetration depth, and also provides a basis for discussion of the alternating current properties of superconductors.

1.3 The Two-fluid Model and Thermal Properties of Superconductors

Early in the study of superconductivity it was observed that there is a discontinuity in the specific heat of a metal at its transition temperature in zero applied magnetic field. This takes the form of a λ-shaped anomaly with temperature (see figure 1.2) for the difference between specific heats of the superconducting and normal states. In zero applied field there is no latent heat associated with the transition, which is thus 'second order' according to the definitions of statistical physics.

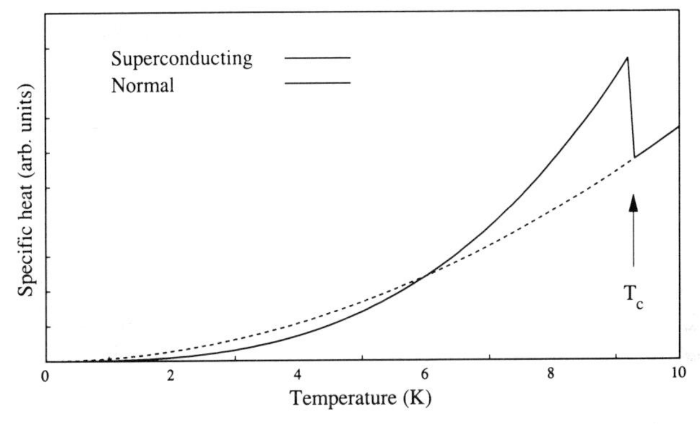

Figure 1.2 Specific heat of niobium in superconducting and normal states

Basic thermodynamic relationships allow the specific heat difference between the two phases to be related to the temperature variation of the critical magnetic field dH_c/dT. Consider the Helmholtz free energy per unit volume $f_s(H)$ of the superconducting phase as a function of applied field H:

$$f_s(H, T) = u - Ts + p - \mu_0 \int m \cdot dH \tag{1.3}$$

where m is the induced magnetisation, s is the entropy density, u is the internal energy density and p is the pressure. This is just a generalisation from the free energy of an ideal gas to that for a substance with non-negligible magnetisation. The term $-\mu_0 \int m \cdot dH$ is analogous to the term

$\int p\,dV$. For a pure material, in the shape of a needle, below the critical field the sample exhibits perfect diamagnetism so that $\boldsymbol{m} = -\boldsymbol{H}$, i.e.

$$f_s(H,T) = u - Ts + p + \mu_0 H^2/2. \tag{1.4}$$

The difference between the free energy of the superconducting state in a field H and in zero field is thus just

$$f_s(H,T) - f_s(0) = \mu_0 H^2/2.$$

The transition from superconducting to normal state happens when the applied field $H = H_c$, corresponding to that value for which the free energies $f_s(H_c)$ and f_n are equal. In other words the free-energy difference between the zero field states at temperature T is simply

$$f_n(T) - f_s(0,T) = \mu_0 {H_c}^2/2 \tag{1.5}$$

where we have assumed that the normal state free energy is field-independent, which is reasonable since superconductors are only very weakly dia- or paramagnetic in the normal state. For each phase the entropy density is just the differential of the Helmholtz free energy with temperature $s = (\partial f/\partial T)$ and the specific heat per unit volume at constant pressure is $c = T\partial s/\partial T$ so that

$$c_n - c_s = T\partial(s_n - s_s)/\partial T = -\mu_0 T[(\partial H_c/\partial T)^2 + H_c(\partial H_c/\partial T)].$$

At $T = T_c$, $H_c = 0$ so that

$$c_n - c_s = -\mu_0 T(\partial H_c/\partial T)^2. \tag{1.6}$$

The difference appears as a positive discontinuity at T_c although it becomes negative as T is reduced. For pure elemental superconductors the temperature dependence of $H_c(T)$ is known to follow a parabolic law of the form

$$H_c(T) = H_0[1 - (T/T_c)^2] \tag{1.7}$$

where H_0 is the critical field value at $T = 0$ K. This expression predicts the shape of the specific heat anomaly in zero applied field quite accurately for most superconductors.

The importance of thermal properties to the development of our subject is that these results introduced the useful phenomenological description of superconductivity provided by Gorter and Casimir's 'two-fluid' model (Casimir 1940). The assumptions are that the current-carrying electrons are divided into two separate but interpenetrating groups, each with strongly differing properties. As the conduction electrons pass through the lattice the 'normal' fraction behaves just like conduction electrons in the normal state but the 'superelectron' fraction has zero entropy and experiences zero resistance. The proportion of the two fractions is temperature-dependent, so that at $T = 0$ K all electrons are in the superfluid state, whereas at $T = T_c$ this fraction falls to zero. The superelectrons effectively short-circuit the normal ones, carrying all the direct current through a superconductor. A metal therefore cannot support a direct voltage across it, with an associated direct current, when it is in the superconducting state. However the entropy of the normal electrons makes itself felt through the thermal properties of superconductors below T_c. The Gorter–Casimir model proposes that the fraction x of superelectrons should vary with T according to

$$x(T) = 1 - (T/T_c)^4. \tag{1.8}$$

In this way it was not only possible to explain qualitative aspects of superconductivity but also quantitative predictions became available, and thus it served as a useful picture until, some 20 years later, a full microscopic explanation appeared.

1.4 Ginzburg–Landau Theory

Much of the contents of the preceding section may be familiar to many readers. It has been included here with the aim of providing a self-contained introduction for newcomers. The basics of what is required for an understanding of large-scale applications of superconductivity (electrical engineering) are very different from what we need for the present subject and have been omitted in order that the summary may be very brief. We now proceed with a discussion of the Ginzburg–Landau (GL) theory, formulated in 1950, a phenomenological model like the two-fluid picture.

However GL theory marks an important break with that simple view. It assumes from the outset that superconductivity represents a state of matter exhibiting *long-range order*, and that one may treat a piece of superconductor as a single quantum entity to which a *complex order parameter* may be everywhere assigned. Even before it was taken up by Ginzburg and Landau, the order parameter concept had already been extensively used in the study of phase transitions (particularly by Landau himself). For example it is useful to define the magnetisation of the domain of a ferromagnet as an order parameter. Below the Curie temperature T_{Curie} the ferromagnet will exhibit a spontaneous magnetisation and the order parameter will have a finite value (which will decrease as T approaches T_{Curie}). Just above this temperature the spontaneous magnetisation (and thus also the order parameter) becomes zero and we say that the phase transition to the non-magnetic state has taken place. The concept may be extended far beyond this simple outline. Particularly, the spatial variation of the order parameter may be of great importance in understanding the phase change itself as well as those practical situations involving physical systems undergoing transitions. This is certainly true, as we shall see below, for small scale superconductivity.

The GL theory starts from the assumption that the superconducting state is characterised by a complex order parameter and our treatment of superconducting electronics is also firmly committed to this approach. It is for this reason that the microscopic Bardeen–Cooper–Schrieffer (BCS) theory (Bardeen *et al* 1957) will not be described in any detail in this book, since in the author's view this approach tends to obscure the macroscopic nature of the superconducting state. There are many excellent texts dealing with the microscopic theory, some of which are included in the references. A brief summary of the BCS theory is given in a later section (1.11) and here and there throughout the book BCS results will be included where needed for our treatment, but only an outline explanation of this microscopic theory will be given.

The fact that, unlike ferromagnetism, a complex order parameter is required for the superconducting state serves to emphasise that we are dealing with quantum mechanics on a macroscopic scale with the latter phenomenon. In the early development of quantum mechanics Schrödinger was obliged to introduce complex wave functions in order to establish the correct relationship between the kinetic energy E and momentum $\boldsymbol{p}$ of a free particle of mass m, as exemplified by the non-relativistic expression $E = p^2/2m$ and the de Broglie relations $E = \hbar\omega$ and $p = \hbar k$. Similarly, in order to derive the magnetic flux exclusion principle and the quantisation of flux we shall find that a complex superconducting order parameter is needed.

1.4.1 Superconductivity and macroscopic quantum order

In a metal at room temperature 10^{23} electrons move through 10^{23} positive ions and, in spite of strong electrostatic and exchange interactions between them, each behaves essentially like an independent free electron, apart from effective mass changes and a certain amount of damping due to electron–phonon or electron–electron interactions. This represents an enormous simplification of what would otherwise appear to be an exceedingly complex many-body problem. The difficulties which physicists faced in understanding the microscopic nature of the superconducting state stemmed in part from a failure to appreciate that the conventional free-electron model of a metal does not necessarily represent the ground state. An appreciation of the different symmetries existing in the superconducting and normal states would doubtless have speeded up the unravelling of the theory of superconductivity.

Over the past 30 years or so an idea which has received much attention is that a *broken symmetry* is associated with any phase transition. For example, the change of phase from liquid to solid is associated with the breaking of translational symmetry, whereas the ferromagnetic transition arises from a loss of rotational invariance. For superconductors a spontaneous symmetry breaking occurs at the transition temperature T_c so that the rather obscure symmetry of *gauge invariance* no longer applies.

1.5 Superconductor Electrodynamics and 'Gauge Invariance'

Conventionally it is possible to add the time and spatial derivatives of an arbitrary function $\chi(r, t)$ to the potentials φ and A describing the electromagnetic fields, without any physical consequences. (This arbitrariness is exploited in the solution of many practical problems for which a particular choice of $\chi(r, t)$ allows the calculations to be much simplified.) Long-range order in the superconducting state makes this 'symmetry' of arbitrary local gauge choice of $\chi(r, t)$ no longer possible. In going from the normal, free-electron, state to the ordered superconducting state of broken symmetry the metal undergoes a phase transition, the new state being described by a complex order parameter

$$\psi(r) = \psi_0(r)\exp(i\varphi(r)) \tag{1.9}$$

which is a function of both position and, in the general case, of time. ψ may be thought of as a wave function describing the macroscopic quantum nature of superconductivity. Above T_c the normal state is the lowest in energy, with $\psi = 0$ everywhere. The ground state of the metal at any temperature is found by minimising the Gibbs free-energy density f, this being the form of free energy required when we wish to ignore the energy cost of setting up the external field. This process shows that at any temperature there will be a certain fraction of the electrons with higher energies, corresponding to normal conduction states. We see here how GL theory explains the two-fluid model of section 1.3. The order parameter $|\psi^2|$ gives an estimate of the fraction of electrons which is in the superconducting state, this fraction being unity at $T = 0$ and reaching zero at the superconducting transition temperature T_c. Over a limited temperature range near T_c the free energy f can be written as a series expansion in ψ with only a finite number of significant terms:

$$f = f_0 + \alpha\psi^2 + (\beta/2)\psi^4 + (1/2m)((-i\hbar\nabla - 2eA)\psi)^2 \tag{1.10}$$

where each coefficient (α, β etc) is temperature-dependent and can be expanded in a power series in $(T - T_c)$. Note that this expression includes a term which depends on the vector potential A. Almost all the topics dealt with in this book are concerned with the response of superconductors to applied electric or magnetic fields. It is thus necessary to establish just how the order parameter responds to such fields. In Appendix A we set out the physical basis for equation (1.10). For those who do not wish to read through this derivation it may be sufficient to point out that the presence of an applied magnetic field adds a contribution to the potential energy of the system as well as the better known contribution from the electric scalar potential. This addition must be subtracted from the total energy operator to yield the *generalised kinetic energy* operator, which is the fourth term on the right-hand side of equation (1.10). Ginzburg and Landau (1950) were the first to write down this expression for the free energy. For a derivation of this 'generalised kinetic energy', the significance of which is not limited to superconductivity, see Appendices A.1 and A.2.

It is straightforward to show that the temperature dependences of the last two terms are

$$\alpha = A(T - T_c)$$
$$\beta = \text{constant.}$$

The solution for ψ is found by minimising the free energy f as ψ is varied. At temperatures sufficiently near T_c only the first term in the power series expansion of $(T - T_c)$ need be retained and this simplification allows the free-energy minimising condition to be written

$$(1/2m)(-i\hbar\nabla + 2eA)^2\psi + \alpha\psi + \beta|\psi^2|\psi = 0. \quad (1.11)$$

This is the Schrödinger equation for the order parameter with the addition of a non-linear term. The equation can be solved for ψ if the boundary conditions are appropriately defined, leading to the important result that there is a characteristic length ξ which describes the range over which the order parameter will significantly change. A more rapid change involves an additional energy cost. The length ξ is related to α by the expression

$$\xi(T) = (\hbar^2/2m\alpha)^{1/2}. \quad (1.12)$$

The current density carried by the superconductor in the presence of a magnetic vector potential A is

$$j = -ie\hbar/m(\psi^*\nabla\psi - \psi\nabla\psi^*) - (4e^2/m)\psi^*\psi A$$

derived by analogy with the quantum mechanical relationship for the current due to a single particle of charge e. In the superconducting case the effective charge on the carriers is taken to be $2e$ in view of the microscopic theory result that paired electrons constitute the supercurrent carriers. Using equation (1.9) for ψ we may deduce that

$$j = (2e\hbar/m)|\psi_0{}^2|\nabla\varphi - (4e^2A/m)|\psi_0{}^2|. \quad (1.13)$$

This relationship between order parameter, current density and vector potential enables the variation of magnetic field within the bulk of the superconductor to be calculated. A simple derivation (see section 1.8) shows that going into the specimen, away from the surface, the magnetic field tends rapidly to zero with a characteristic length known as the GL penetration depth

$$\lambda = (m\beta/4e^2\mu_0\alpha)^{1/2}. \quad (1.14)$$

This behaviour immediately explains 'flux exclusion' (the Meissner effect) which showed that when a metal is cooled through its transition temperature in a magnetic field the field is expelled from the bulk of the specimen, rather than being trapped in it. For reasonably small applied fields the ground state is the flux-excluded one, this state being achieved by the screening effect arising from supercurrents flowing in a layer of thickness λ on the surface of the specimens. If a sufficiently large magnetic field is applied the Gibbs free energy of the normal state becomes lower than that of the superconducting state so that normal properties are restored.

1.6 A Macroscopic Quantum View of the Meissner Effect

The model outlined above is capable of describing the properties of pure metals near T_c, including temperature dependence of H_c and λ. One of the major successes of GL theory was to explain the much more complex magnetic properties of alloy systems (see section 1.8). Having accepted the GL notion of a complex order parameter describing the superconducting state, a number of consequences follow. It is most interesting to consider the effect of an applied magnetic field on ψ. This field may be derived from a vector potential A, and its effect on a wave function for a single particle of charge e and mass m is to multiply the complex order parameter by a factor $\exp(2\mathrm{i}eA/\hbar)$, leaving the amplitude unchanged (see Appendix A.2). The order parameter, like any wave function, must be single-valued, so that the line integral of the phase change around any closed contour must be equal to $2n\pi$, where n is an integer. Consider a single piece of superconductor without any holes through it (see figure 1.3). There are an infinite number of closed path integrations which can be performed, starting from, and returning to, the point O. In the limit there is a path of vanishing length, for which the total phase change must clearly be zero. An infinitesimal increase in path length cannot produce a phase change of 2π, such as would be needed to make a transition to a different state. By this continuity argument we see that for all paths entirely within the superconductor $\delta\varphi = 0$. The above expression for the variation of φ with A gives for $\delta\varphi$ around a closed path on which the current density is everywhere zero

$$\delta\varphi = (2e/\hbar)\int A \cdot \mathrm{d}l = 0. \tag{1.15}$$

Stokes' theorem allows the line integral of A to be transformed to a surface integral of $\nabla \times A$ over the surface bounded by the closed path. By definition $\nabla \times A$ is the magnetic field $\boldsymbol{B}$ itself, and the surface integral of $\boldsymbol{B}$ is just the magnetic flux through the surface

$$\int A \cdot \mathrm{d}l = \int \nabla \times A \cdot \mathrm{d}S = \int \boldsymbol{B} \cdot \mathrm{d}S = \Phi = 0. \tag{1.16}$$

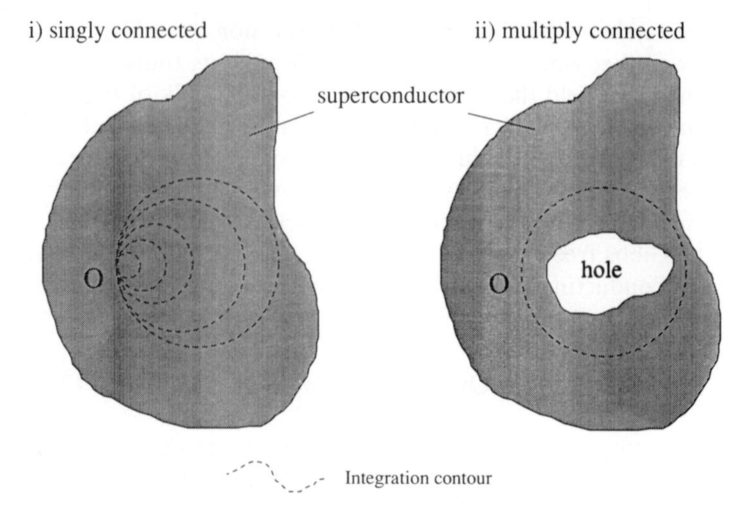

Figure 1.3 Single and multiply connected superconductors

Thus we see that the flux through any contour entirely within a single piece of superconductor is zero, as evidenced experimentally by the Meissner effect.

1.7 Multiply Connected Superconductors

If instead of a single piece we allow our superconductor to have a hole through it, the situation becomes very different. There is in this situation no zero-length line contour surrounding the hole so that the continuity argument cannot be applied. Thus the solution with a total phase change of $2n\pi$ with $n > 0$ cannot be ruled out. In this case equation (1.15) becomes

$$\delta\varphi = (2e/\hbar)\Phi = 2n\pi$$

so that

$$\Phi = nh/2e$$

i.e. the flux through the tube is quantised in units of $\Phi_0 = h/2e$. At first sight this might seem paradoxical since magnetic flux may generally be treated as a classical variable, capable of smooth changes between any limits. In ordinary situations this is true; the current through a solenoid may take on any value we choose. In the case of the hole through the superconductor we must use a self-consistency argument. We know that to screen out magnetic fields from the interior of a bulk superconductor, surface currents must flow. In the same way surface currents flow around the inner surface of the hole and it is the sum of the externally applied flux

Φ_x plus this internally generated flux which must be quantised:

$$\Phi_i = \Phi_x + Li_s = n\Phi_0$$

where L is the ring inductance around which the supercurrent i_s flows. Thus the ring of superconductor responds to any change in external flux $\Delta\Phi_x$ by setting up an equal but opposite flux. Provided that the supercurrents flowing on the surface do not exceed the critical current density for that specimen then, for as long as the specimen remains superconducting, the total flux linking the ring will remain constant, and quantised, at the same value. London predicted flux quantisation in 1950, some six years before a microscopic theory was available, just on the basis that superconductivity represented a macroscopic ordered state. He suggested that the unit of quantisation would be h/e, just twice what it was found to be from the first experimental measurements of Deaver and Fairbank (1961). This result was as predicted by the BCS theory which showed that electron pairs with effective charge $2e$ are the basic supercurrent carriers.

The fact that the allowed states of the ring exhibit quantised flux is sufficient to tell us that the Gibbs free energy of the ring will be periodic in flux, the periodicity being Φ_0, the flux quantum. Thus the free energy of a ring, expressed as a function of both Φ_x and Φ_i, exhibits a series of equally spaced potential wells, whose depth depends on the critical current of the ring. (This periodicity may be demonstrated more fundamentally by solving equation (1.11) numerically, to show that the effective potential in this non-linear Schrödinger equation is periodic.) Figure 1.4 shows the free-energy surface in such a case for a strongly coupled superconducting ring. The local energy minima centred on each quantised value of internal flux are very deep, of order $\mu_0 j_c^2 r^5$ where j_c is the critical current density and r is the ring radius. Only by applying a huge external flux ($\sim \mu_0 r^3 j_c$) can a transition from one internal flux state to another, of lower energy, be brought about. This is only achieved by the destruction of superconductivity in the material so that the order parameter goes everywhere to zero before re-establishing itself at a new value of internal magnetic flux. In view of the unknown dynamics of the transition from the superconducting to the normal, then back to the superconducting state again, the amount of flux admitted to the ring under such circumstances is quite unpredictable.

As the external flux applied to the ring is changed the absolute depth of each minimum also changes, and at any instant the state for which the circulating current is a minimum represents the true ground state. The ring will, however, remain in its initial internal flux state unless the critical applied flux is exceeded, even when this has a higher energy than the lowest state and is hence only metastable.

The lifetime of a metastable state against thermal activation or other decay processes may be qualitatively estimated. If the well depth is ΔU, the critical current density j_c and L the ring inductance, then the lifetime t

against thermal activation to a lower flux state is

$$t \sim [\omega \exp(-\Delta U/kT)]^{-1} \tag{1.17}$$

where ω is an attempt frequency, that is the rate at which the state of the ring oscillates in the potential well which the free-energy surface defines. Near the bottom of the well the shape is, to a good approximation, parabolic. In this case the attempt frequency is just the natural resonant frequency of the ring multiplied by a factor $(2\pi i_c L/\Phi_0)^{-1}$. Inserting typical values for a small ring of $i = 10^{-4}$ A and $L = 1$ nH at a temperature $T = 4$ K, equation (1.17) predicts that the lifetime of a metastable state is of order 10^{22} s, greater than the age of the universe. We have assumed a perfect crystal lattice. In practice imperfections and impurities would presumably shorten this time, but experimentally the expression is verified for low critical currents. For bigger rings with large critical currents a lower limit to the lifetime has been shown by experiment to be at least 10^7 years (Gallop 1976, 1977). Thus for all effective purposes the lifetime of persistent currents is infinite, as is the conductivity, at least for direct currents. This question is considered further in section 1.12.

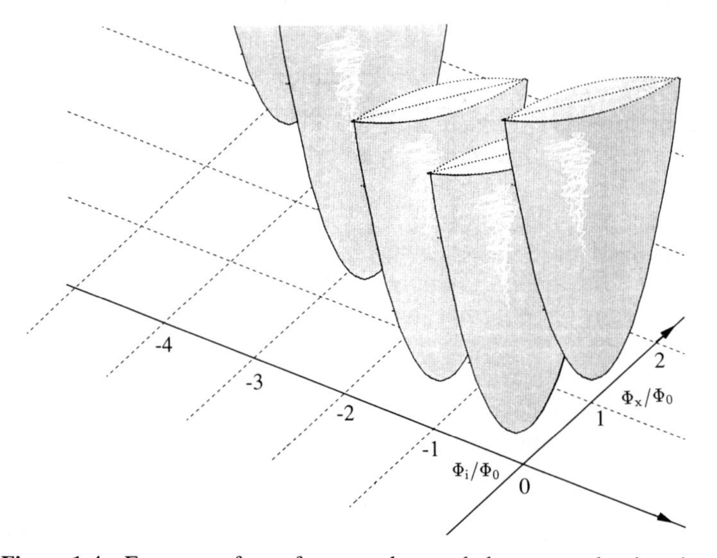

Figure 1.4 Energy surface of a strongly coupled superconducting ring

1.8 Further Insights Using GL Theory

In equation (1.11), the solution of which determines the spatial variation of the order parameter, there is a term $(1/2m)(-i\hbar\nabla)^2\psi$ which represents the kinetic energy of the current-carrying particles. It also demonstrates that an energy price is paid if the order parameter is to change over a short distance, since $\nabla\psi$ will be large in this case. Here is the physical basis for the existence of a coherence length ξ which sets the scale on which ψ can change.

In a large piece of superconductor, far from its boundary on the scale of ξ the order parameter ψ will be constant in space, so that $\nabla\psi = 0$. Substituting this in equation (1.13) gives a simple relationship between the current density $\boldsymbol{j}$ and the vector potential $\boldsymbol{A}$

$$\boldsymbol{j} = -(4e^2\alpha/m\beta)\boldsymbol{A}. \tag{1.18}$$

Since there is also a Maxwell equation linking current and magnetic field $\boldsymbol{B}$, namely

$$\nabla \times \boldsymbol{B} = \mu_0\boldsymbol{j} \tag{1.19}$$

these two may be combined to show that

$$\nabla^2\boldsymbol{B} = (4e^2\mu_0\alpha/m\beta)\boldsymbol{B}. \tag{1.20}$$

The solution of this equation for a plane geometry reveals that B is attenuated exponentially with distance into the superconductor, the length scale being the penetration depth λ, defined above in equation (1.14). (This is yet another demonstration of the Meissner effect, based on the electrodynamics of superconductors summarised by equation (1.18).)

GL theory predicts that ξ has the same temperature dependence as the penetration depth λ, but this is not supported by experiment. This should not surprise us when we recall that the theory is only valid very close to the transition temperature since it involves a series expansion approximation, and retains only the first two terms. To summarise, finite penetration of the magnetic field is a result of the finite coherence length which in turn arises from the inertia, and thus the kinetic energy, of the electron pairs responsible for superconductivity.

1.9 AC Properties of Superconductors

This inertia is also responsible for the ohmic losses in superconductors in applied fields at non-zero frequencies. The electrons' inertia produces a delay between application of an electromagnetic field and the response of pairs to it. The phase lag indicates that pairs cannot respond rapidly enough to screen out totally a time-varying electric field in the surface layer, allowing normal electrons to be accelerated by the unscreened field and then scattered off phonons, other normal excitations or lattice defects and impurities. Thus superconductors exhibit a finite surface resistance to alternating fields, the surface resistance R_s being a strong function of frequency (since this affects the phase lag between current and applied field) and also temperature (since this affects the number of superconducting excitations at the expense of normal excitations). A higher density of pairs improves the screening, fewer normal electrons lead to fewer scattering possibilities; thus both effects combine to reduce dissipation as T falls. The dependence of $R_s(\omega, T)$ has been shown to follow the phenomenological expression

$$R_s(\omega, T) \propto (\omega^2/T)\exp(-\Delta/kT). \tag{1.21}$$

As with many seemingly straightforward analytical results in superconductivity, equation (1.21) has only a limited range of applicability, when $T < T_c/2$ and for very pure single-crystal specimens, but its intrinsic simplicity generates considerable physical insight.

1.10 Superconductivity of Alloys

The ratio of the two characteristic lengths in superconductivity (the coherence length ξ and the penetration depth λ) $\varkappa = \lambda/\xi$ has particular significance for those materials on which the practical applications of superconductivity are based. One of the GL theory's outstanding successes was in explaining the existence of 'type II' superconductivity in alloys. For these materials, even in an idealised situation and geometry, the Meissner effect does not exist above the lower critical field H_{c1}, although superconductivity, as demonstrated by zero resistance, persists to some much higher limiting field H_{c2}. (This is in contrast to 'type I' materials, usually pure metals, for which only the lower critical field exists, so if the applied field exceeds this the material reverts to normal metallic behaviour.)

If $\varkappa$ is greater than $1/\sqrt{2}$ the surface energy of the superconducting–normal interface becomes negative, indicating that the interior of a superconductor in an applied field greater than H_{c1} will gain energy by allowing the field to penetrate, along tubes of normal metal parallel to the applied field. On a classical picture this subdivision might be expected to continue to an infinite density of infinitely thin tubes. However a quantum limit is reached when a single flux quantum is associated with each normal tube. The interaction energy of the normal tubes (known as vortices) leads to mutual repulsion, ensuring that in a perfectly crystalline solid they would take up a regular array, equally spaced on a triangular lattice, whose spacing is a function of applied field. In real superconductors vortex lines can become pinned to lattice imperfections or impurities, so that the perfect vortex lattice will be likewise deformed. Most of superconducting electronics is concerned with low-field limits where for all technically useful superconductors the applied fields are less than the lower critical field H_{c1}. However it will be necessary to discuss the properties of type II superconductivity when we consider the fabrication and uses of superconducting thin films (Chapter 6) and the use of superconductors for electromagnetic screening (Chapter 7).

1.11 The BCS Theory of Superconductivity

The microscopic theory of superconductivity formulated by Bardeen, Cooper and Schrieffer (BCS) in 1957 represented the culmination of almost 50 years of work into understanding the origins of superconductivity. This theory has been only occasionally referred to until this point in our introduction. As was suggested above, the Ginzburg–Landau theory is more appropriate to a treatment of superconductivity which stresses its macroscopic quantum nature, even though its range of validity and its quantitative predictions are far more limited. However this section will include a brief description of the elements of BCS microscopic theory, giving special emphasis to one or two results which will be needed in later treatments of the Josephson effects. It should also be pointed out that Gor'kov (1959) showed that the microscopic theory reduces to GL theory close to T_c.

1.11.1 The indirect electron–electron attraction

Very early in the study of superconductivity the existence of persistent currents led to the conclusion that the phenomenon represented some collective state of the conduction electrons in a metal. A mechanism leading to such a state was hard to imagine since its existence seems to imply an attraction between electrons. The Coulomb repulsion of like charges was expected to dominate inter-electron coupling. An important clue was furnished by the observation that the transition temperatures for different isotopes of a particular element varied as $M^{-1/2}$, where M is the isotopic mass (this relationship is now known to be only rather approximately obeyed in general). The discovery suggested that in some way collective excitations of the lattice (phonons) might mediate an attractive coupling between conduction electrons. Frolich (1950) proposed an exchange of virtual phonons between electrons which could, under favourable circumstances, lead to a net attractive force. A few years later Cooper (1956) calculated that this interaction would operate between a pair of electrons, effectively binding them together under appropriate conditions. For two electrons with momenta $\boldsymbol{k}$ and $\boldsymbol{k}'$, exchanging a phonon with momentum vector $\boldsymbol{q}$, the interaction energy is

$$V(\boldsymbol{k},\boldsymbol{k}',\boldsymbol{q}) = g^2\hbar\omega/\{[E(\boldsymbol{k}+\boldsymbol{q}) - E(\boldsymbol{k})]^2 - (\hbar\omega)^2\}$$

where $E(\boldsymbol{k})$ is the kinetic energy of an electron with momentum $\hbar\boldsymbol{k}$, $\hbar\omega$ is the phonon energy and g is the coupling constant between the electron and phonon.

This interaction energy $V(\boldsymbol{k}, \boldsymbol{k}', \boldsymbol{q})$ can be positive, implying an attraction between electrons, provided that

$$|E(\boldsymbol{k}) - E(\boldsymbol{k}')| < \hbar\omega.$$

Cooper made a number of simplifying assumptions when calculating the total coupling energy of a pair, taking the summation over $\boldsymbol{k}$ and $\boldsymbol{k}'$ into account as well as the repulsive Coulomb interaction, and was able to show that

$$V \sim g^2/\omega_0$$

where $\hbar\omega_0$ is the Debye energy for the metal, and there is only a significant coupling if $|\boldsymbol{k} - \boldsymbol{k}'| < \omega_0$. Note that $\hbar\omega_0 \ll E_0$, the Fermi energy of the metal, so that effectively an electron is paired with another having equal but opposite momentum.

It was left to BCS to incorporate the basic pairing interaction into a full microscopic theory of superconductivity. The Pauli exclusion principle, governing the behaviour of the Fermi sea of conduction electrons, also had to be included. Assuming a mean value of V for the virtual phonon exchange energy, independent of $\boldsymbol{k}$ and $\boldsymbol{k}'$, and that only electrons above the Fermi sea can interact via this mechanism, it was straightforward to show that the average binding energy of a pair was

$$E \sim 2\hbar\omega_0 \exp(-2/N(0)V)$$

where $N(0)$ is the density of states of the conduction electrons at the Fermi surface. Provided that $E > kT$, it is energetically favourable for pairs to form.

1.11.2 The superconducting energy gap

It then proved possible to calculate the energy difference between the state in which pairs had formed and the normal metal ground state, it being $W(0)$ at $T = 0$ K where

$$W(0) = -\ N(0)\hbar\omega_0 E$$

since $N(0)\hbar\omega_0$ is an estimate of the number of pairs within an appropriate energy range of the Fermi level. There is thus an energy gap Δ between superconducting and normal states. Raising the temperature provides more thermal energy and this tends to disrupt the coupling between pairs, thus reducing their number. Consequently Δ also falls with increasing T, reaching zero at T_c. Here is the BCS justification for the two-fluid model (section 1.3). The 'superelectron' fraction is represented by the pairs, which are responsible for dissipationless current flow. Pairs may be split by external influences other than temperature, such as the absorption of high-frequency photons or phonons. To a good approximation, an incident photon of energy $\hbar\omega = 2\Delta = 3.5kT_c$ is sufficient to destroy superconductivity by dissociating a pair, even at $T = 0$ K. We shall meet this process again when considering the high-frequency limit of the Josephson effect. The experimentally determined variation of $\Delta(T)$ can be quite accurately approximated by a simple analytical expression

$$\Delta(T)/\Delta(0) = [\cos(2\pi(T/2T_c)^2)]^{1/2}$$

and this dependence is plotted in figure 1.5. The BCS model provided a simple explanation of the far-infrared absorption measurements made during the 1950s which showed that superconductors exhibited significant dissipation at these frequencies.

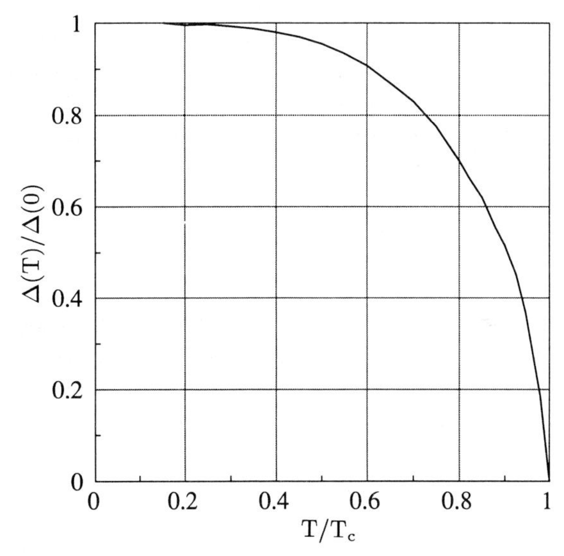

Figure 1.5 Temperature dependence of the energy gap Δ

Even before the BCS theory provided a very full microscopic explanation of superconductivity, some experimental results on thermoelectric effects and high-frequency properties of superconductors had suggested that there was a temperature-dependent energy gap separating the ground state from an excited normal conduction state. The energy gap concept has been developed far beyond the simple idea outlined here. It can be shown that if the gap is treated as a complex parameter which varies with space and time it can be identified with the order parameter of the GL theory.

At the time of writing an electron-pairing mechanism is known to operate in the high-temperature superconducting cuprate materials. However this is not expected to be phonon-mediated. Nevertheless these materials exhibit properties which make it clear that an energy gap with similar temperature dependence to that of a BCS superconductor still exists, albeit with a rather larger value than the weak coupling model would suggest.

1.12 The Lifetime of Persistent Currents

The question is often asked 'Is the resistivity of superconductors truly zero for direct currents?'. The answer must be rather more involved than might be expected. Again consider a closed loop of superconducting wire. A state carrying a circulating current can only be metastable. The lifetime of a persistent current flowing in a wire of a given diameter, made from a specified material and operated at a specified temperature, can be theoretically calculated from the coupling energy of the superconducting state and the thermodynamics of thermal fluctuations, which in principle are capable of bringing about transitions corresponding to a decay of the circulating supercurrent. Little (Edelsack 1973) carried out calculations of this type and the results are summarised in table 1.1. It is clear that except for wires much smaller in diameter than the coherence length, or for macroscopic wires within a few microkelvin of T_c, the half-life for the decay of the supercurrent is long compared with the age of the universe. Thus for all practical purposes the resistivity of most superconductors to direct current is zero. (The current flowing in a ring of inductance L would be expected to decay by discrete steps $\Delta i = (2\Phi_0/L)^{1/2}$.) The concept of a finite resistivity implies dissipation proceeding by a continuous loss of stored energy and may be regarded as inapplicable to such a 'lumpy' process.

Table 1.1 Computed lifetimes for two Sn wires, one of 100 Å diameter and the other of 360 Å

Lifetime	100 Å wire (K)	360 Å wire (K)
~ 10^{-14} s	3.7	3.7
10^{-7} s	2.0	3.4
10^{-4} s	1.5	3.26
1 s	1.25	3.10
10 days	1.0	3.06
10^6 years	0.75	2.90

From another viewpoint the question is rather like asking 'Can all the molecules of a gas in a closed vessel move to one end of it at the same time?'. Although the probability of this happening is finite, in principle no-one would waste their time looking for such events in macroscopic gas containers. The lifetime of persistent currents is a question of thermodynamics and so does not really fall into the same category as more fundamental questions like 'Does the photon have a rest mass?', for which the answer is that current theories of QED suggest that its mass is identically zero. As outlined in section 1.9 the resistivity of superconductors for alternating currents is both finite and experimentally measurable. In all real cases in the radio-frequency range and at temperatures below T_c the measured resistivity is greater than equation (1.21) predicts. Considerable effort has been expended in trying to understand the source of this excess dissipation since the Q-factor of superconducting resonators is correspondingly lower than might be hoped for. It appears that surface finish and material purity are important parameters in determining losses, as is the pinning of magnetic flux lines through the resonator walls (Hartwig and Passow 1972).

2

The Josephson Effects

2.1 Introduction

That superconductivity is a macroscopic quantum phenomenon is undoubtedly the main message to be drawn from the previous introductory chapter. There we saw that flux quantisation is a remarkable demonstration of the strange behaviour exhibited by macroscopic systems which are governed by fundamental quantum laws. In this chapter we introduce the Josephson effects, the second piece of fundamental physics which has been exploited in the development of superconducting electronics. If anything they provide an even clearer view of the quantum nature of superconductivity, through a decidedly esoteric mix of properties.

The Josephson effects arise when two pieces of superconductor are separated by a physically small region (on the scale of either the superconducting coherence length ξ or the penetration depth λ), in which the superconductivity is weakened. Such a weakly superconducting region, known as a *Josephson junction* (Josephson 1964), can be produced in a variety of ways which will not be outlined until section 2.5. As in the previous chapter, we wish first to introduce the fundamental concepts underlying the superconducting properties before expanding on the basics with a discussion of more detailed properties of specific structures.

2.1.1 The effect of a weak link on the internal energy of a superconducting ring

A key result from the previous chapter is that the internal energy U of a superconducting ring is a periodic function of the applied magnetic flux Φ_x. In this chapter we will be concerned with what happens to $U(\Phi_x)$ as the critical current of a short length is reduced from the value appropriate to the rest of the ring. For a uniform ring made from a homogeneous superconductor at a temperature well below T_c, adjacent quantised flux states are separated by high energy barriers (see figure 1.4), and it is only by applying a very large external field to the ring that the internal flux state of the ring can be altered. Flux changes brought about in this way are unpredictable and catastrophic. The height of the energy barrier separating adjacent flux states of the ring is of order

$$U = Li_1^2/2 \tag{2.1}$$

(where i_1 is the critical current), whereas the energy change associated with a change of one flux quantum in the applied flux is

$$U = \Phi_0^2/2L. \tag{2.2}$$

One may surmise that smooth changes between adjacent internal flux states may be brought about by relatively small changes in external flux only when the first of these energies is comparable with, or less than, the second, leading to the condition

$$i_1 \leqslant \Phi_0/L. \tag{2.3}$$

The critical current of a short length of the ring might be most obviously lowered by reducing the cross-sectional area of a part of the ring. The bulk critical current density for a typical type I superconductor well below its transition temperature is of order $10^7\ \mathrm{A\,m^{-2}}$. In order that any flux state should be well defined with a long lifetime, the energy difference between adjacent flux states must be much greater than the energy kT of a typical thermal fluctuation:

$$\Phi_0^2/2L \gg kT.$$

This imposes the condition that at say 4.2 K, the inductance must be of order 1 nH or less. It is difficult to produce a bulk ring with $L < 10^{-11}$ H so these two conditions taken together require $10^{-11} < L < 10^{-9}$ H, and combining this with equation (2.3) we see that the critical current i_1 of the constricted region must be $\leqslant 10\ \mu\mathrm{A}$, limiting its cross-sectional area to $<(1\ \mu\mathrm{m})^2$. A number of methods of producing weak-link regions possessing the appropriate critical current have been developed and these will be described in section 2.8.

2.2 The DC Josephson Effect

In the weak-link region it is no longer possible to find a path through it on which the supercurrent density j is everywhere zero. (This is perhaps the most general definition of a weak link.) The importance of this fact can be seen by rewriting equation (1.13) in the following form:

$$\nabla\varphi = [mj/(2e\hbar\,|\,\psi^2\,|\,)] + 2eA/\hbar. \tag{1.13}$$

The non-zero value of j introduces an additional contribution φ' to the phase change of the order parameter in the weak-link region. Integration of equation (1.13) from one side of the link to the other, as shown in figure 2.1, gives

$$\varphi' = \frac{m \int_1^2 \boldsymbol{j} \cdot \mathrm{d}\boldsymbol{l}}{e\hbar \mid \psi^2 \mid}. \tag{2.4}$$

where the magnetic flux in the weak-link region is assumed to be negligible. The linear relationship between φ' and j of equation (2.4) only applies in the low-current limit when $\mid \psi^2 \mid$ is equal to its bulk material value. For higher currents $\mid \psi^2 \mid$ reduces to zero smoothly, although the details of this need not concern us here. The low-current regime yields the constant of proportionality, k, between φ' and the total current and is

$$k = ml/2Se\hbar \mid \psi^2 \mid \tag{2.5}$$

where l is the length and S is the cross-sectional area of the weak link. The free-energy surface $U(\Phi_x, \Phi)$ can also give information on the effect of current on φ'. The circulating supercurrent is related to U via the expression

$$i_s = \mathrm{d}U/\mathrm{d}\Phi_x. \tag{2.6}$$

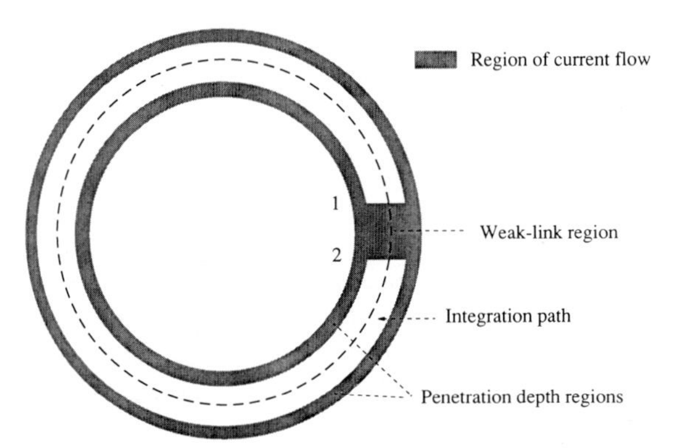

Figure 2.1 Superconducting ring interrupted by weak link

Since U is a periodic even function of the included flux Φ_i within the ring, so too is i_s. The current-induced phase change φ' must be included in the single-valued condition on the order parameter, which becomes

$$\varphi' + (2e/\hbar) \int \boldsymbol{B} \cdot \mathrm{d}\boldsymbol{S} = \varphi' + 2\pi\Phi_i/\Phi_0 = \varphi' + 2\pi(\Phi_x + Li_s)/\Phi_0 = 2n\pi. \tag{2.7}$$

The periodic nature of $i_s(\Phi_i)$ means that quite generally it can be expanded as a Fourier series in (Φ_i/Φ_0), an odd function in view of the form of equation (2.6). In the limit of very weak coupling it seems reasonable that the magnitude of the terms in the Fourier expansion decreases very rapidly with increasing order. So in this limit we can replace the series by just its lowest-order term

$$i_s = i_1 \sin(2\pi\Phi_i/\Phi_0) = i_1 \sin[2\pi(n - \Phi_i/\Phi_0)] = i \sin(\varphi') \qquad (2.8)$$

where $i_1 = k^{-1} = 2e\hbar n_s/m$ represents an estimate of the maximum supercurrent which can pass through the link, corresponding to a phase difference across it of $\varphi' = \pi/2$. In this expression we have introduced a normalisation condition on the order parameter, that $|\psi^2| = n_s$, the local pair density. Equation (2.8) is the first of the well known Josephson effect relationships (Josephson 1962).

2.2.1 The AC Josephson effect

The second important Josephson effect can be derived by considering the same hypothetical ring arrangement if we now imagine that a time-varying external flux is applied to the ring. In addition to the single-valued order parameter condition we must also apply Faraday's law, requiring that the time rate of change of flux linking the ring is equal to the voltage developed around the ring. Thus if the flux is increased linearly with time,

$$V = -\,\mathrm{d}\Phi_x/\mathrm{d}t = \text{a constant}, \qquad (2.9)$$

the voltage developed is independent of time, and it appears only across the weak link. The time-varying flux will also influence the circulating supercurrent through its effect on the phase change across the weak link. Using the first Josephson equation it is clear that

$$i_s = i_1 \sin[2\pi(Vt + Li_s)/\Phi_0]. \qquad (2.10)$$

In other words the macroscopic quantum nature of superconductivity requires that a constant voltage drop generates an alternating supercurrent at a frequency f proportional to the applied voltage, $f = 2eV/h$. Equivalently, combining equations (2.8) and (2.10) we see that a constant voltage causes a linear increase of φ with time:

$$\mathrm{d}\varphi/\mathrm{d}t = 2eV/\hbar. \qquad (2.11)$$

The constant of proportionality, $2e/h$, is the inverse of the quantum of magnetic flux and has the value 483.594 THz V^{-1}.

We may generalize this ring arrangement to consider the voltage drop across a weak link joining two separate pieces of superconductor. This requires an additional assumption that the phase of the order parameter in one region is not directly affected by the phase far away; in other words we have a 'local' theory of superconductivity. The assumption seems to be well justified in practice. The two Josephson relationships were originally derived for the specific case of two superconductors separated by a weak link through which electron pairs could tunnel. However, the effects are observed for any form of weak coupling, whether via thin superconducting bridges, normal metal links or by tunnelling through semiconductors as well as insulators. One of the remarkable features of the Josephson effects is the extreme accuracy of equation (2.11). Many attempts, both theoretical and experimental, have been made to discover corrections to its very simple form. To date none have been found, provided certain size and temperature conditions are satisfied, some tests having demonstrated a precision as high as 1 in 10^{16}. This topic will be discussed at greater length in section 2.4.1.

2.2.2 Current–voltage characteristics of a Josephson junction

It would be helpful in understanding the behaviour of superconducting weak links to be able to describe their properties by means of an electrical equivalent circuit. The derivation of equations (2.8) and (2.11), which embody the essence of the Josephson effects, was made by starting from some fundamental macroscopic quantum features of superconductivity. However, this treatment, due to Bloch (1970), cannot describe the full equivalent circuit of a Josephson junction. For our present purpose, in order to understand most applications of the Josephson effects and SQUIDs, it is sufficient to describe a junction by the equivalent circuit of figure 2.2. If the device is biased from an adjustable constant current source, and the bias current i is increased from zero, initially there will be no voltage drop across the junction, although the passage of a current through the Josephson element will introduce a phase difference across it, given by equation (2.8). When the bias current exceeds the junction critical current a voltage will appear and the phase difference will become time-dependent, but in addition both normal and displacement currents will now flow through the parallel conduction paths formed by R and C respectively. Experimentally it is found that point-contact junctions have approximately ohmic resistance above the critical current but the situation is different when the coupling is by means of electron-pair tunnelling between two insulated pieces of superconductor. There will still be an effective dissipative channel associated with the passage of quasi-particles (effectively normal state electrons) which shunts the Josephson element, but this is now very non-ohmic, varying very strongly with voltage. The microscopic nature of the dissipative processes is not well understood and further work in this area is required. For the present discussion an ohmic resistance will be assumed for the normal electron channel.

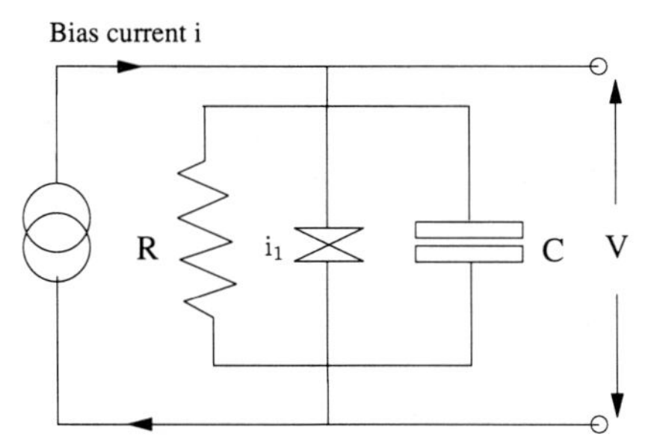

Figure 2.2 Schematic diagram of resistively shunted junction

The geometry of a thin-film Josephson tunnel junction is determined by the requirement that the two superconductors must be very close to allow pair-tunnelling to occur (see section 2.5). For an insulating barrier the thickness must be only of order 1–2 nm. This means that the two superconducting planes will have significant capacitance and this cannot be ignored in the equivalent circuit. For point-contact junctions the capacitance is much smaller. In this case it might be expected that the geometry would result in a series inductive term but in practice this seems not to be necessary when modelling their behaviour. This topic will be discussed in more detail in sections 2.5 and 2.6, when specific junction types and methods of preparation will be dealt with. To determine the DC current–voltage characteristic of a junction described by the circuit of figure 2.2 we first note that the bias current is the sum of three terms:

$$i = i_1 \sin \varphi + V/R + C\, \mathrm{d}V/\mathrm{d}t. \qquad (2.12)$$

However the second Josephson relationship $V = (\hbar/2e)\, \mathrm{d}\varphi/\mathrm{d}t$ (equation (2.11)) relates the voltage drop across the device to the time rate of change of φ so that this equation becomes a second-order non-linear differential equation for φ:

$$i = i_1 \sin \varphi + (\hbar/2eR)\, \mathrm{d}\varphi/\mathrm{d}t + (\hbar C/2e)\, \mathrm{d}^2\varphi/\mathrm{d}t^2. \qquad (2.13)$$

To determine the current–voltage characteristic of the junction we must solve for the time-average value of the voltage or, equivalently, $\langle \mathrm{d}\varphi/\mathrm{d}t \rangle$. If $C = 0$, which is a reasonable first approximation for a point-contact junction, or even for a small-area tunnel junction, equation (2.13) can be solved directly by integration to give

$$\begin{aligned} V &= 0 & i &< i_1 \\ V &= i_1 R[(i/i_1)^2 - 1]^{1/2} & i &> i_1. \end{aligned} \qquad (2.14)$$

Thus if the time-average voltage is plotted as a function of the bias current it clearly has a parabolic dependence as shown in figure 2.3(a). This behaviour is reasonably well followed in practice for the types of junction specified above, particularly for junctions with rather small values of shunt resistance ($R < 1\ \Omega$).

If the junction capacitance is not negligible, equation (2.13) can be solved numerically and it is then found that the current–voltage characteristic may exhibit 'hysteresis', so that the path taken for increasing i is not the same as that for decreasing i. Figure 2.3(b) illustrates this behaviour. The effect occurs when the condition

$$\beta_c = 2\pi i_1 R^2 C / \Phi_0 > 1 \qquad (2.15)$$

is satisfied, and β_c is known as the hysteresis parameter.

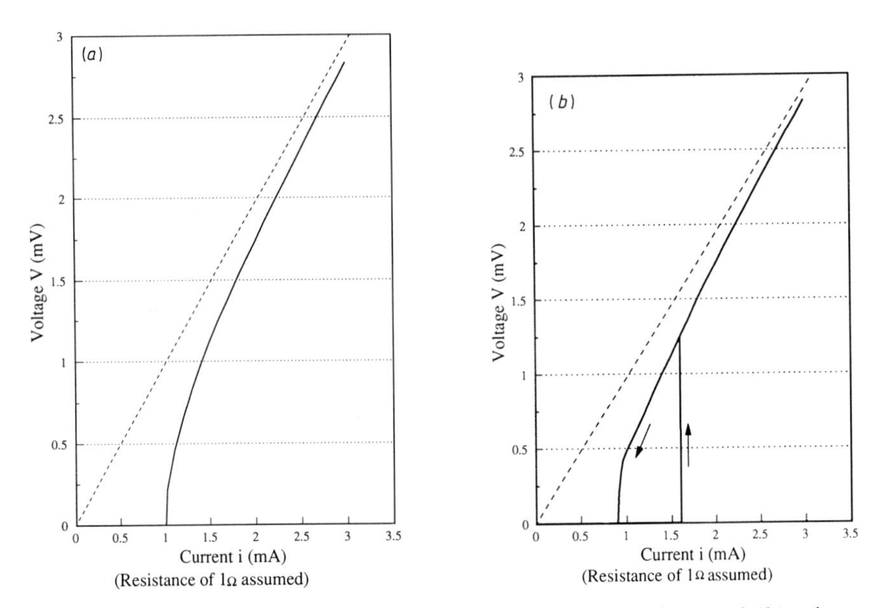

Figure 2.3 *I–V* characteristic of (a) a low-capacitance junction, and (b) a junction with larger C

2.2.3 Quantum interference in wide junctions in a magnetic field

Up to this point it has been assumed that the dimensions of the Josephson junction in the plane orthogonal to the current flow are quite negligible, so that the supercurrent density is uniform throughout, as is the phase change of the order parameter across the junction. In the case of a large Josephson junction formed between superconducting electrodes, as shown schematically in figure 2.4, this assumption is no longer valid. We have already seen in Chapter 1 and Appendix A.2 that a magnetic field influences the order parameter phase θ, and this applies to the electrodes as well as to the weak-link region. To be explicit, equation (1.13) shows that in going from a to b within the superconducting electrode the phase change due to a magnetic vector potential $\boldsymbol{A}$ is

$$\theta_b - \theta_a = 2e/\hbar \int \boldsymbol{A} \cdot \mathrm{d}\boldsymbol{s}.$$

This expression applies when a and b are chosen to be far from the junction region so that in equation (1.13) $\boldsymbol{j} = 0$ along the entire path. Consider a magnetic field $\boldsymbol{B}$ applied in the z direction to a junction, of width w in the x direction, perpendicular to $\boldsymbol{B}$. Let us assume that the phase change from one side of the junction to the other (from a to d) at $x = 0$ is $\delta\theta(0)$. Then the junction phase difference at any point x may be found by integrating from a to b and from c to d as shown in figure 2.4. The additional phase change becomes

$$\delta\theta(x) - \delta\theta(0) = (2e/\hbar)\left(\int_a^b \boldsymbol{A} \cdot \mathrm{d}\boldsymbol{s} - \int_c^d \boldsymbol{A} \cdot \mathrm{d}\boldsymbol{s}\right) = (2e/\hbar) \oint \boldsymbol{A} \cdot \mathrm{d}\boldsymbol{s} \quad (2.16)$$

where the two line integrals have been turned into a closed-loop integral by adding the very short segments from d to a and from b to c. The contributions may be made identically zero since, without loss of generality, we

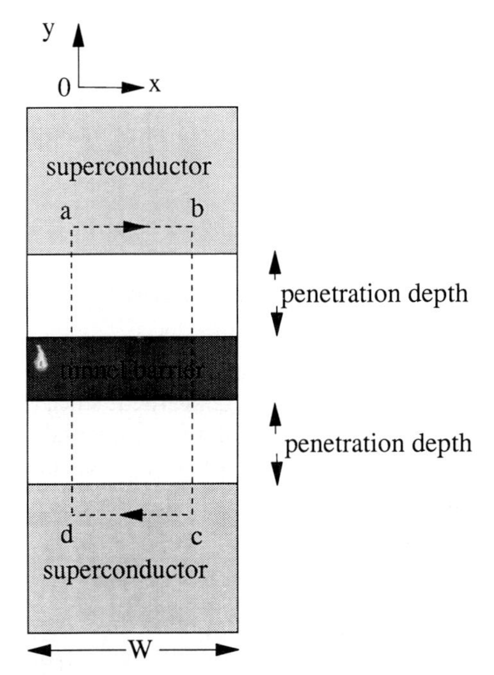

Figure 2.4 Schematic of Josephson junction with integration path through barrier

may choose a gauge in which $A_y = 0$ (see Appendix A.3.1). Applying Stokes' theorem, as in equation (1.16), it is clear that the difference in junction phase change between any two points along the x axis is

$$\delta\theta(x_1) - \delta\theta(x_2) = 2e\Phi/\hbar = 2\pi\Phi/\Phi_0$$

where Φ is the magnetic flux within the junction between two planes which have x values x_1 and x_2.

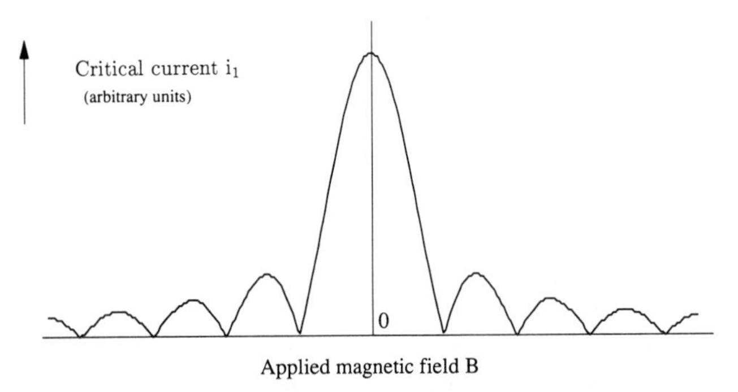

Figure 2.5 Dependence of large-area tunnel junction critical current on applied magnetic field (sin(AB)/AB)

The DC Josephson effect (equation (2.8)) requires that the supercurrent density $j(x)$ at any point in a wide junction is equal to $j_1 \sin(\delta\theta(x))$, where j_1 is the critical current density. To find the total supercurrent i which flows in the presence of a magnetic field one need only integrate this expression across the junction:

$$i = \int j_1 \sin \delta\theta(x) \, \mathrm{d}x = \int j_1 \sin(2\pi\Phi(x)x/\Phi_0 + \delta\theta(0)) \, \mathrm{d}x$$

$$= i_1 [\sin \delta\theta(0) \sin(2\pi\Phi/\Phi_0)] / (\pi\Phi/\Phi_0) \qquad (2.17)$$

and Φ is now the total flux contained within the junction. For an insulating barrier of thickness t the effective area of the junction is approximately $w(t + 2\lambda)$ as far as the magnetic field is concerned, since the field enters the electrodes to a distance of the order of the superconducting penetration depth λ (see section 1.4.1). The effect of an applied field is to modulate periodically the maximum supercurrent which can flow through a wide Josephson junction. The expression

$$i_{\max} = \sin(2\pi\Phi/\Phi_0)/(2\pi\Phi/\Phi_0)$$

is plotted in figure 2.5 as a function of applied flux. At each null point of the supercurrent the total change in phase difference from one edge of the junction to the other is $2n\pi$ where n is an integer. For these supercurrent null-points the current density is not everywhere zero but changes sign n

times across the junction width. Recent experiments in which the local supercurrent density has been probed with a fine laser beam (Lhota *et al* 1984) confirm the predicted distribution (see figure 2.6). Incidentally, the similarity of figure 2.5 to the intensity of light diffracted onto a screen by a slit was responsible for this phenomenon being called 'quantum interference' and thus for the acronym 'SQUID'.

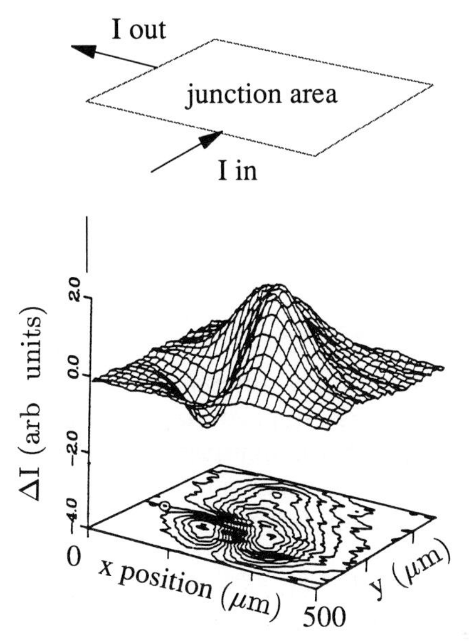

Figure 2.6 Measured Josephson current distribution in Pb–oxide–Pb crossed films tunnel junction. The current feed is shown schematically at the top, with the current distribution shown in contour form below (see Lhota *et al* 1984)

2.2.4 Quantum interference when the Josephson current is not negligible

The current flowing through the junction also generates a magnetic flux density $B'(x)$ which so far we have ignored. Reverting to the simplified notation of section 2.1 we set $\varphi(x) = \delta\theta(x)$, the phase change across the junction at x. Assuming that the applied magnetic field is zero in this case and differentiating equation (2.16) twice with respect to x yields the relationship

$$\partial^2\varphi/\partial x^2 = 2\pi(t + 2\lambda)(\partial B'(x)/\partial x)/\Phi_0 \qquad (2.18)$$

between φ and B'. Maxwell's fourth equation simplifies to

$$\nabla \times \boldsymbol{B} = \mu_0 \boldsymbol{j}(x)$$

if we consider a stationary state so that $dE/dt = 0$. Furthermore if the junction is assumed to be infinitely long in the z direction so that all z derivatives will be zero, a further simplification results:

$$\partial \boldsymbol{B}'/\partial x = \mu_0 \boldsymbol{j}. \tag{2.19}$$

Substituting this in equation (2.17) yields

$$\partial^2 \varphi/\partial x^2 = 2\pi(t+2\lambda)\mu_0 j_1 (\sin \varphi)/\Phi_0 \tag{2.20}$$

where the DC Josephson relationship has also been used. For small values of φ the sin φ term may be linearised and the resulting second-order differential equation has simple exponentially decaying solutions from the edge of the junction towards the centre. By symmetry

$$\varphi(x) = \varphi(-x)$$

and

$$\varphi(|x|) = \varphi(|w/2|)\exp[(w/2 - |x|)/\lambda_J].$$

In this linear approximation the Josephson current has a similar spatial dependence, screening out the applied magnetic field from the interior of the junction. This is strongly reminiscent of the Meissner effect in bulk superconductors and the self-screening length in the junction case is known as the Josephson penetration depth:

$$\lambda_J = \{(\Phi_0/[2\pi\mu_0 j_1(t+2\lambda)]\}^{1/2}.$$

Typical parameters yield λ_J values of around 1 mm, some three to four orders of magnitude greater than a typical penetration depth for a bulk material.

2.2.5 *Time dependence of the phase difference*

In the discussion above, the order parameter phase difference has been assumed to be constant in time. Relaxing this constraint implies, through the AC Josephson relationship (equation (2.11)), that a voltage (and thus an electric field E_y) exists between the two electrodes. This can be accommodated by adding the displacement current term to equation (2.19):

$$d\boldsymbol{B}/dx = \mu_0 \boldsymbol{j} + dE_y/dt$$

where μ_0 is the assumed magnetic permeability of the junction region. The electric field extends over a distance t through the junction so the voltage drop $V = E_y t$. Combining this with the AC Josephson relationship and substituting into equation (2.20) yields a non-linear wave equation for the spatial and temporal variations of φ:

$$\partial^2\varphi/\partial x^2 - (1/c'^2)\partial^2\varphi/\partial t^2 = (\sin\varphi)/\lambda_J^2 \tag{2.21}$$

where c' is the electromagnetic wave velocity in the junction and is related to the free space velocity c as follows:

$$c' = c[t/2\varepsilon(t + 2\lambda)]^{1/2}.$$

For most junction geometries and materials $t \ll \lambda$ so that

$$c' \sim c(t/2\varepsilon\lambda)^{1/2} \sim c/10.$$

When the junction width is sufficiently small that variations of φ in the x direction may be ignored, equation (2.21) reduces, in the small-amplitude limit, to

$$(-1/c'^2)\partial^2\varphi/\partial t^2 = \varphi^2/\lambda_J^2$$

This has sinusoidally varying solutions with frequency $\omega_J = c'/\lambda_J$, representing the interchange of stored energy between electrostatic charging energy and the 'kinetic' energy associated with supercurrent flow through the junction. By analogy with the behaviour of ionised gases this excitation mode is known as the Josephson plasma oscillation. As an alternative to the above microscopic picture we may express ω_J in terms of lumped circuit parameters. For a junction of cross-sectional area A the capacitance $C = \varepsilon\varepsilon_0 A/\mathrm{t}$ whereas the critical current $i_1 = j_1 A$ so that

$$\omega_J = (L_J C)^{-1/2} = (2ei_1/\hbar C)^{1/2}$$

where $L_J = \hbar/2ei_1$ is an effective inductance associated with the Josephson coupling energy of the weak link and ε is the permittivity of the tunnel barrier.

For junctions of larger width the application of a magnetic field B in the z direction will alter i_1 as outlined above in section 2.2.1 so that ω_J varies with B. Similarly a direct current bias $i < i_1$ gives rise to the possibility of small oscillations in phase difference about a mean value of $\varphi > 0$. The solutions of equation (2.21) then require that ω_J varies also with bias current i. Detailed experimental measurements of these magnetic field and bias-current-dependent effects on the plasma modes of junctions have been reported (Pedersen *et al* 1972).

2.3 The AC Properties of Josephson Junctions

The prediction by Josephson of an alternating supercurrent through a junction when it is biased into the finite voltage regime prompts the question 'How might this current be detected?'. Associated with an oscillating current there are time-varying electric and magnetic fields which give rise to electromagnetic radiation, emitted by the junction. No sharp cut-off exists for the maximum frequency at which the supercurrent oscillates. It is found that the AC supercurrent is attenuated in amplitude at bias voltages above about $2\Delta/e$, where 2Δ is the energy gap (section 1.11.1). The energy of the internally generated photons is great enough in this regime to disrupt electron pairs, reducing the amplitude of the critical current. For example, for niobium $2\Delta/e = 2.7 \times 10^{-3}$ V, corresponding to a frequency of 1.3 THz. At first sight the voltage-biased Josephson junction appears very attractive as a microwave source, tunable over a very wide frequency range by adjustment of the direct voltage bias. However there is only a very limited amount of available radiated power. An upper-limit estimate of this power may be simply given. Suppose that the junction has a critical current of 1 mA, typical of a reasonably large-area tunnel junction, and that it is biased at a voltage of a few tens of microvolts, corresponding to a supercurrent frequency in the microwave region. The total dissipation will then be of the order of tens of nanowatts. Not all of this power will appear in the form of electromagnetic radiation since a large fraction of the total will be dissipated in the junction and its immediate surroundings as heat. However this rough calculation yields an upper-limit estimate, showing that only very low radiated powers can be expected. The real situation is rather worse since for most junction configurations, particularly for large tunnel devices, the characteristic impedances of the structures are very low, giving a bad mismatch between the oscillator and free space. The difference between c' and c (section 2.2.5) is a major cause of this mismatch which further severely limits the amount of power radiated. The output power is sufficient to make it straightforward to detect the emitted radiation with a fairly sensitive receiver.

The radiated power level can be considerably increased by matching the junction impedance to that of free space, or a waveguide, by means of a microwave transformer, but a narrow-band device results, so that the intrinsic advantage of broad-band tunability has been lost.

There is a second serious limitation, arising from the linewidth of the emitted radiation. For a junction at temperature T, having a normal state shunt resistance R, there will be Johnson noise voltage fluctuations appearing across it which will add to the steady bias voltage, frequency modulating the alternating supercurrent. The mean square voltage fluctuation is

$$\langle V^2 \rangle = 4kTR \, \mathrm{d}f \tag{2.22}$$

where $\mathrm{d}f$ is the linewidth being considered, which also enters into the frequency modulation expression:

$$\mathrm{d}f = \langle V^2 \rangle^{1/2} / \Phi_0. \tag{2.23}$$

Combining equations (2.22) and (2.23), and solving for df gives

$$\mathrm{d}f = 4kTR/\Phi_0^2. \tag{2.24}$$

(This relationship will appear again in Chapter 5, in a more detailed form when noise thermometry is discussed.) For a junction operating at the normal boiling point of liquid helium, 4.2 K, with a typical resistance of 1 Ω the linewidth is of order 1 MHz, not a very impressive figure for a microwave source.

The linewidth df may be very significantly reduced by coupling the junction to a high-Q resonant cavity (Elsley and Sievers 1974). Equation (2.24) still determines the linewidth, as the strongly coupled cavity will reduce the differential resistance of the junction, particularly in the voltage-bias region where the internally generated frequency is close to the cavity resonant frequency, or a harmonic of it. In this way the linewidth can be reduced to 1 kHz or less. However, just as for the impedance-matching case, this use of a narrow-band resonant structure sacrifices the most attractive feature of broad-band voltage tunability.

2.3.1 Radiation generation by arrays of junctions

One possible way to increase the emitted power is to couple together a number of junctions, biased at the same voltage. For n junctions radiating incoherently the power will just be multiplied n times. However, if the n junctions can be persuaded to radiate in phase the field amplitudes add, and since the power is proportional to the square of the field amplitudes, the total power output will be n^2 times that from a single junction. There are at least two mechanisms which can bring an array of junctions into phase coherence. First they may be coupled via the electromagnetic field in free space surrounding the superconductors. Unless the size of the array is much smaller than the free space wavelength of the electromagnetic radiation there will be complications due to spatial phase shifts between different parts of the array. Alternatively the junctions may be brought into coherence by direct coupling of the order parameter in the superconducting regions themselves. For this situation complications are introduced if the array is more than one-dimensional, since then there are multiple current paths joining any pair of junctions. The mechanism responsible for the latter form of coupling is not well understood. Thus, following an initial flurry of activity and confidence, the work in this area has gone back to attempting to understand fully the nature of the processes going on in the simplest form of array, that is of just two junctions. An alternative approach uses the motion of a soliton in the form of an isolated magnetic flux vortex trapped in a long ($\sim$1 mm) Josephson tunnel junction, which oscillates back and forth along the junction. This recent work will be outlined in Chapter 6, when actual tunnel junction fabrication techniques and circuits will be dealt with.

2.4 The Influence of Applied Electromagnetic Radiation on Josephson Junctions

The low radiated power level from a voltage-biased Josephson junction prevents this technique from providing a simple means of demonstrating the existence of the AC supercurrent. Instead an indirect method can most readily be used, and historically too this one was the first to give convincing evidence of the correctness of Josephson's predictions. The method involves irradiating a junction with microwaves. The effect on the DC current–voltage characteristic is dramatic: constant-voltage 'steps' are induced, equally spaced in voltage with a separation which is directly proportional to the microwave frequency applied. The mechanism which produces the steps involves the phase locking of the alternating supercurrent to the applied microwave frequency, or a harmonic of it, over a range of bias current. The current width of a step corresponds to a frequency-pulling range for the junction. Frequency pulling, resulting in phase locking, can happen whenever the internally generated frequency is reasonably close to the microwave frequency, or any harmonic of it. Thus the voltage spacing of the steps is

$$\Delta V = hf/2e \tag{2.25}$$

where f is the applied microwave frequency. The current amplitude of these induced steps is a complicated function of the applied power, with the steps corresponding to each harmonic of the applied frequency oscillating in amplitude as the power is increased, out of step with one another. If one can assume a voltage-biased junction (not a realistic assumption in most applications) an analytical solution for the step amplitudes is readily obtained, as follows.

In the case of a constant voltage bias V, the phase difference of the order parameter across the junction evolves linearly with time, according to the following relationship:

$$\varphi(t) = 2eVt/\hbar = 2\pi Vt/\Phi_0. \tag{2.26}$$

The supercurrent flowing through the junction is just

$$i(t) = i_1 \sin \varphi(t). \tag{2.27}$$

The addition of an alternating microwave voltage $V_1 \sin 2\pi ft$ also influences the time evolution of the phase via equation (2.26), and so affects the supercurrent. The complete expression for $i(t)$ becomes

$$i(t) = i_1 \sin[(2\pi t/\Phi_0)(V + V_1 \sin 2\pi ft)]. \tag{2.28}$$

We are interested in any time-independent term in this current since that component will show up in the DC current–voltage characteristic. To examine this question, it is simple to expand the above relationship in terms of an infinite series of Bessel functions $J_n(V_1/\Phi_0 f)$:

$$\sin[(Vt/\Phi_0) + (V_1 t/\Phi_0)\sin(2\pi ft)]$$
$$= \sum_{n=-\infty}^{\infty} J_n(V_1/\Phi_0 f)\sin[2\pi(V/\Phi_0 - nf)t + \theta]. \quad (2.29)$$

From this expression it is clear that whenever $V = n\Phi_0 f$ the argument of the sine function becomes time-independent, and the supercurrent has a DC component whose magnitude varies with the adjustable phase angle θ. This can take values between $+\pi/2$ and $-\pi/2$, so that for each value of n the range of additional direct supercurrent which is allowed to flow becomes $\pm i_1 J_n(V_1/\Phi_0 f)$. The current–voltage characteristic in this (admittedly unrealistic) voltage-biased example would consist of a series of spikes of amplitude given by this expression, at voltages equally spaced by $\Phi_0 f$. Figure 2.7(a) shows how the spike amplitudes would vary with applied microwave voltage amplitude V_1 for some low-order values of n. In most practical situations the microwave and direct bias sources will have high impedances compared with that of the junction itself and so a model employing current sources must be used. No simple analytical solution exists in this case but numerical solutions can be carried out simply. As before, the I–V characteristic is modified by the addition of current 'steps' at constant voltage, again equally spaced by amount $\Delta V = hf/2e$ (see figure 2.8(b)). However although the width of the steps shows the same qualitative dependence on microwave applied power as in the voltage-bias case, with the amplitudes oscillating with increasing power, the widths of the steps are more rapidly attenuated as the power is increased, and also the dependence on f is more marked (see figure 2.7(b)).

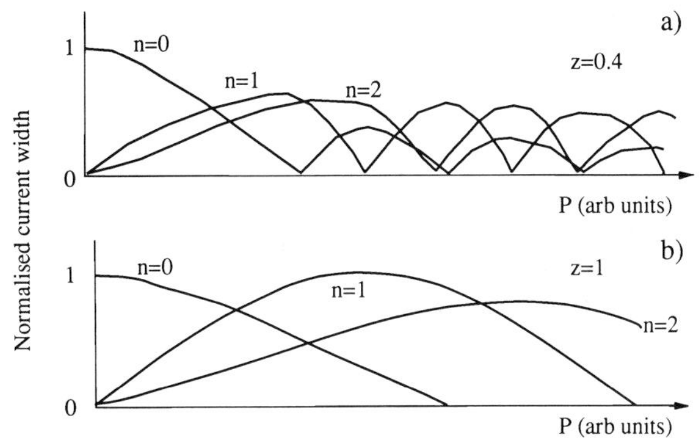

Figure 2.7 Dependence of Shapiro step width ($n = 0, 1, 2$) on microwave power level P for two different values of normalised microwave frequency $z = \hbar f/2ei_cR$: (a) $z = 0.4$; (b) $z = 1$

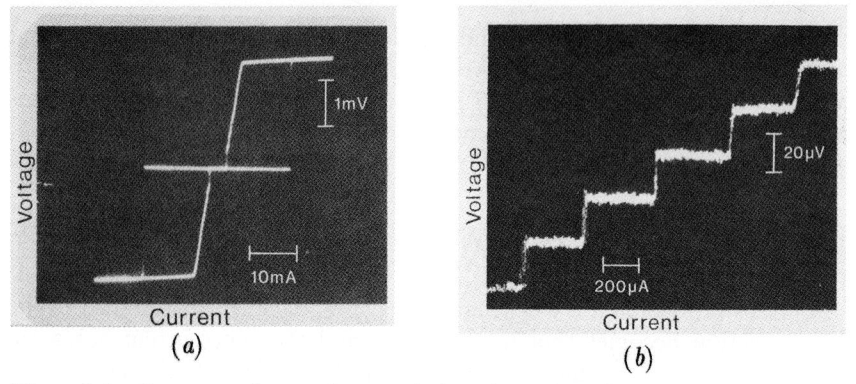

Figure 2.8 Current–voltage characteristic of a Josephson tunnel junction: (*a*) *I–V* plot for a Pb–PbO_x–Pb tunnel junction; and (*b*) part of the same characteristic when 10 GHz microwave radiation is applied (Hartland 1981)

2.4.1 Measurement of h/2e

The presence of constant voltage steps in the current–voltage characteristic of a microwave-irradiated Josephson junction was the first demonstration of the correctness of Josephson's predictions concerning the AC properties of weakly coupled superconductors (Shapiro 1963). The same arrangement was used for the first practical application of the effects, to a measurement of the fundamental physical constant $h/2e$. Soon after the first demonstration of induced steps, workers at the University of Pennsylvania undertook a thorough investigation of the properties of Josephson junctions (Parker *et al* 1969). One of the questions which first concerned them was 'how accurate is the relationship between bias voltage and supercurrent frequency, as embodied in equation (2.11)?' It became clear from their experiments that this relationship, although extremely simple, was extremely accurately obeyed. They were unable to find any systematic corrections depending on material, temperature, frequency or magnetic field, given the limits of their voltage measuring technique. Thus the Pennsylvania group proposed that the system provided an accurate method for determining the value of the flux quantum $h/2e$. The method is simple in principle. The voltage difference V separating a large number of steps is determined and the frequency f of the microwave source measured. If the number of step intervals measured is m, then $h/2e = V/mf$. The accuracy achieved in the first series of measurements was about two parts per million (PPM) and the stage was rapidly reached where the limitation on precision was set by the ability to measure rather small voltage differences (a few mV). Experimentally it became clear that, because of the fundamental but simple nature of the AC Josephson effect, it could provide a quantum standard of voltage which would be more stable and accurate than the then standard, the Weston cell. By 1971 a number of national standards laboratories were using the effect to monitor their conventional units of voltage, and following an international agreement in 1975 the maintained standard of voltage has been defined in terms of the Josephson effects and a fundamental constant of physics, $h/2e$, the flux quantum (Φ_0). This topic will be discussed in more detail in Chapter 8, including later realisations of Josephson voltage standards and experimental limits on the accuracy of equation (2.11).

2.4.2 *Radiation detection using Josephson junctions*

Josephson junctions have found other applications as well. Incident microwave radiation at low power levels depresses the amplitude of the zero-voltage step, i.e. the ordinary DC supercurrent of the junction. (This can be seen from the J_0 term in equation (2.29).) This forms the basis of a sensitive broad-band video detector. Furthermore if two microwave or far-infrared signals are incident on a junction in the finite voltage regime both will influence the alternating supercurrent. The extremely non-linear properties of the junction, exemplified by equation (2.29), mix the incident frequencies very effectively. This arrangement forms the basis of two applications: the first as a narrow-band heterodyne detector with high sensitivity. The second is as a high harmonic mixer, which allows frequency intercomparison in a single stage between a microwave source and an infrared source such as an HCN laser. Sources up to a frequency of 3 THz have been measured and two frequencies differing by a ratio of 1000 have been intercompared in this way (Blaney and Knight 1974). More details of both these applications will be given in Chapter 6.

2.5 Real Junction Structures

2.5.1 *Pair tunnelling structures*

The original theoretical derivation of the Josephson effects (Josephson 1962) was formulated on the assumption that the weak-link structure connecting the two superconductors was a thin insulating barrier through which electron pairs could 'tunnel'. Tunnelling is a purely quantum mechanical effect, quite forbidden in classical physics, which results from the finite value of the wave function describing a particle in a region in which the particle is classically forbidden, on account of its negative total energy (Landau and Lifshitz 1972). Close to an allowed–forbidden interface the wave function will decay into the negative energy region with a characteristic length set by $d = 1/k = \hbar/[2m(V-E)]^{1/2}$, where V is the potential energy and E is the total energy of the particle, whose mass is m. In any region over which V is constant the wave function Ψ varies spatially as $\exp(ik.r)$ so that provided the kinetic energy is positive ($E - V > 0$), k is real and the wave function shows oscillatory wave behaviour. If, on the other hand, $E - V < 0$ then k will be purely imaginary and Ψ will be exponentially damped as r increases into the forbidden region. Imagine now that another region of positive energy exists on the far side of a narrow forbidden region (figure 2.9). On the right-hand side, solutions for Ψ will again be travelling waves with a real value of k. Boundary conditions require that at the two interfaces both Ψ and its spatial derivative are continuous. Thus for a forbidden region of finite width sandwiched between two allowed regions the wave function has finite travelling solutions on either side. The probability of finding the particle at any point r is proportional to $|\Psi(r)^2|$. Thus if the particle is incident on the left-hand side of the barrier, there is a finite probability that it will be transmitted through the barrier, rather than being reflected. In the limit of low tunnelling probability the fraction of incident particles which are transmitted becomes

$$P = 16(E/V)\exp\{-2[2m(V-E)/\hbar]^{1/2}t\}$$

where t is the barrier thickness.

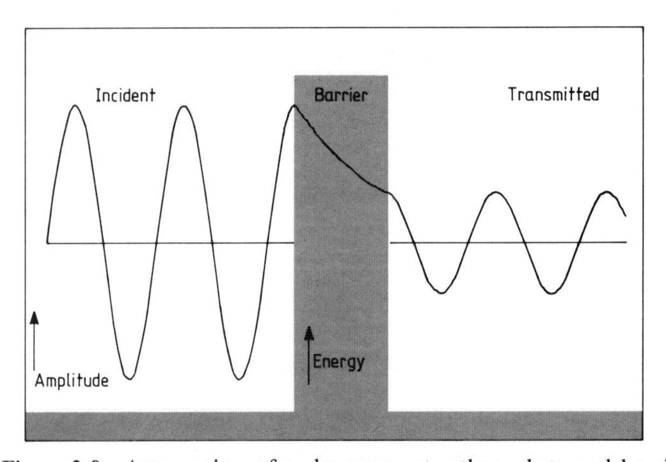

Figure 2.9 Attenuation of order parameter through tunnel barrier

To find the current flowing through a tunnel barrier we need to know n, the rate at which particles are incident on the barrier, as well as P. The current is just $nP\,\mathrm{s}^{-1}$. Our simple calculation of the current of free particles tunnelling through a barrier can be made specific. Ambegaokar and Baratoff (1963) have described the more realistic situation of a BCS state Josephson tunnel junction with variable temperatures. In this case we must sum over all the matrix elements connecting pair states on either side of the barrier. Very approximately, for an ideal rectangular barrier at $T=0$ the critical current density of the junction is

$$J_c \sim 2\pi^2(e^2/\hbar)\Omega_1\Omega_2\langle T^2\rangle\Delta \tag{2.30}$$

where Ω_1 and Ω_2 are the pair densities of states on either side and $\langle T\rangle$ is a typical matrix element connecting states across the junction. $\langle T^2\rangle$ is exponentially damped with V and t, just as P is in the simple model above. Equation (2.30) can be further developed to show that

$$J_c = \pi\Delta/2er$$

where r is the tunnelling resistance per unit area of the junction when its electrodes are in the normal state, an empirical parameter which includes the properties of the junction such as Ω_1 and Ω_2, V and t which are difficult to calculate. For a superconductor such as lead and a barrier height $V = 10$ eV, typical of an insulator, a barrier thickness of order 1 nm will provide critical current densities in the range 10^6–10^7A m^{-2}, which spans the useful range for tunnel junctions. Such a barrier, only a few molecular layers thick, can be set up by mechanically positioning two discrete pieces of superconductor relative to one another (this method forms the practical basis of the superconducting scanning tunnelling microscope). However most practical devices require that an insulating layer a nanometre thick is put down on a first superconducting surface, before the second superconductor is deposited, without disrupting the insulator, a difficult problem whose solution depends on the materials used. The most widely used of the techniques that have been developed to produce Josephson junctions will be dealt with in more detail in Chapter 6. The need to vacuum deposit the insulating and second superconducting layers has dictated that Josephson tunnelling junctions are essentially planar structures.

2.5.2 *Junctions with non-insulating barriers*

It is not only insulating barriers which can be used in tunnel junctions. Semiconductors differ from insulators in that the lowest allowed conduction states are much closer to the filled valence band, so that at room temperature at least a few of the electrons have sufficient thermal energy to occupy these states and contribute to a measurable conductivity. Since the barrier height is lower, junctions with semiconducting barriers have higher critical current densities than insulating barrier junctions of the same thickness. Alternatively the same current density can be achieved with a much thicker barrier.

Perhaps even more surprising is that the 'tunnel barrier' can be a normal metal. Electrons are not able to exist in equilibrium as Cooper pairs in the normal metal region so a barrier still exists, although it is extremely low, of order 10 meV. There is the further complication arising from the proximity effect. This is briefly summarised by stating that near a normal metal–superconductor (N–S) interface superconductivity is weakened in the S region and also a narrow section of the N region is made superconducting, due to the finite range over which pairs entering the normal metal travel before breaking up. Thus the barrier is not contiguous with the region of normal metal and its height is not constant across its thickness. In addition normal electrons in the superconductor can flow freely through the 'barrier' region to the other side so that the junction is 'shorted' by a normal electron channel with a resistance of only perhaps 1 $\mu\Omega$.

2.5.3 The Riedel peak and the cos φ term

Our discussion of the Josephson effects in the finite voltage regime has assumed that the junction critical current has the same value when alternating currents flow as it has in the direct Josephson effect, regardless of the direct voltage across the junction. In other words we have assumed that i_1 is frequency-independent. In considering the transfer of superconducting electron pairs (see section 1.11) through the junction by tunnelling we have also ignored the presence at finite temperatures of 'quasi-particle excitations' in the superconductors. These, which may generally be regarded as the normal electron component of the two-fluid model (see section 1.3), can also tunnel through the barrier. They are responsible for the ohmic behaviour of the I–V characteristic in the finite voltage regime. When $V > 0$ it is necessary to consider the density of states of both pair and quasi-particle distributions as a function of energy since, recalling section 1.11, the tunnelling rate and hence the current depends on the density of occupied states on one side of the barrier multiplied by the density of unoccupied ones on the other. Figure 2.10 indicates the density of states for pairs and normal electrons in a superconductor in the region of the Fermi level, showing the superconducting energy gap. For energies near the lower or upper bounds of the superconducting energy gap the pair density of states diverges. Thus it is not surprising that the maximum pair current i_1 also diverges when the voltage separating the two superconductors is equal to 2Δ. This was first pointed out by Riedel and has been further analysed by many authors. For a review of the present state of knowledge see Barone and Paterno (1981). Quasi-particle tunnelling is likewise voltage-dependent so that a conductivity $\sigma_1(V)$ may be assigned. For a perfect tunnel junction $\sigma_1(V)$ will be almost equal to zero for $V < 2\Delta$ but for higher voltages it assumes a more or less constant value.

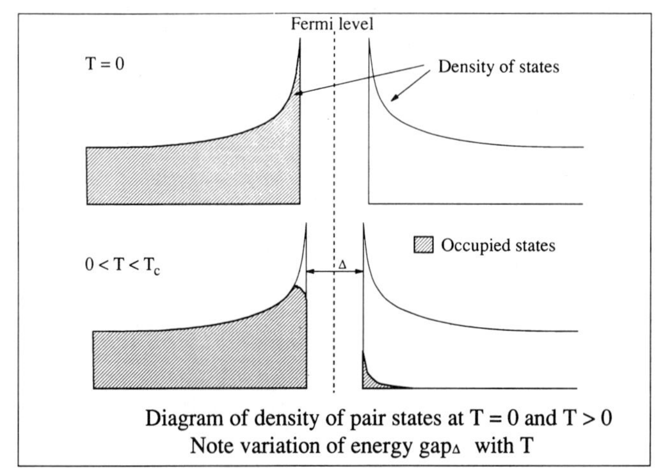

Figure 2.10 Density of states distribution around the Fermi level of a superconductor showing the energy gap

There exists an even more subtle voltage-dependent term in the expression for the total current through a Josephson tunnel junction. This is the so-called quasi-particle pair interference term. As well as simple elastic or inelastic pair transfer through the junction, more complex processes are also possible. The most probable process involves the destruction of a pair in one electrode and the creation of one in the other electrode, together with the transfer of a quasi-particle. This involves interference effects between the wave functions on either side of the barrier, resulting in a contribution to the quasi-particle current which depends on the cosine of the phase difference φ. Including both quasi-particle components, the total junction current becomes:

$$i = i_1(V,T)\sin\varphi + (\sigma_1(V,T) + \sigma_2(V,T)\cos\varphi)V. \tag{2.31}$$

The presence of a $\cos\varphi$ term has been confirmed in a number of experiments. However theory and experiment cannot yet be said to be in agreement since the sign of the term differs between prediction and what is observed. We will not discuss further these sophisticated modifications of the Josephson current since they are beyond the scope of this treatment and have yet to provide practical applications.

2.6 The Physics of Point-contact Junctions and Microbridges

2.6.1 Point-contact junctions

Tunnel junctions were the first type of Josephson structure to be proposed, are by far the best understood theoretically and, in view of the sophisticated fabrication techniques now available, are becoming of dominant practical importance. However they are not the only structures exhibiting the full range of macroscopic quantum effects. Point-contact junctions still have advantages for specific applications. This type of weak link is strongly reminiscent of 'cat's whisker' diode detectors. A sharpened superconducting point is pressed lightly against a flat superconducting 'anvil'. The contact region then exhibits all of the basic Josephson effects which have been discussed above. A schematic view is shown in figure 2.11. Although qualitatively similar, the I–V characteristics of point-contact junctions differ in detail from those of tunnel devices, the major contrast being that the resistance is often essentially voltage-independent for point contacts and there is rarely pronounced structure in the region of the gap voltage (contrast figures 2.3(*a*) and 6.3). Hysteresis effects are only infrequently seen, particularly for freshly made point-contact surfaces, and the modulation of critical current by a magnetic field requires high values of field to be detectable. Both of these observations indicate that the contact area (and hence the junction capacitance) are small compared with fabricated tunnel barriers.

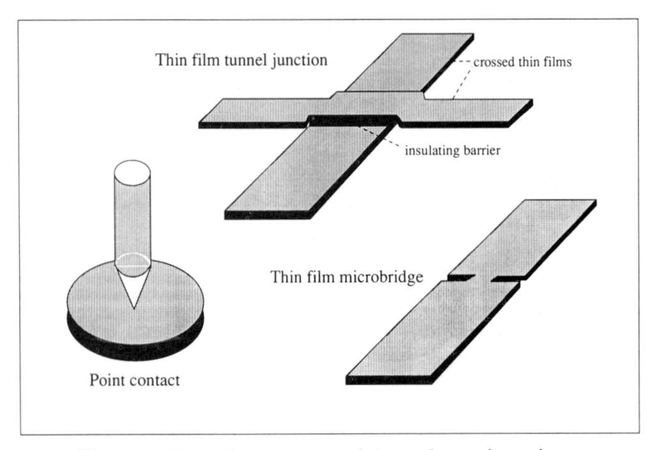

Figure 2.11 Three types of Josephson junction

2.6.2 Microbridge weak links

The third general type of Josephson junction is formed where a very small constriction is produced in a thin superconducting film. Although superficially similar in nature to a point-contact junction, in fact the properties of microbridges are significantly different. The shunt resistance of the planar device is found to be smaller and more temperature-dependent than that of a point contact. Constrictions which are small in all three dimensions, on the scale of the temperature-dependent coherence length $\xi(T)$, exhibit the whole range of Josephson effects, including sinusoidal current–phase relationships. As with point contacts, structure at the gap voltage, or multiples of it, is not usually very pronounced. When the microbridge is long compared with $\xi(T)$ in any direction, further complicating features appear in the I–V characteristic.

2.6.3 The resistively shunted junction model

A simple phenomenological theory, the resistively shunted junction (RSJ) model, has been developed to describe the properties of point contacts and planar microbridges. This model has proved very successful in describing most of the Josephson junctions used up until the present day, but it should be borne in mind that in terms of a microscopic theory very little is known of the operation of any type of weak link in which the superconducting channel is shunted by a significant quasi-particle-carrying channel. This whole topic belongs in the realm of *non-equilibrium superconductivity*. We shall not discuss it further here, but only describe qualitatively what is believed to occur in point-contact and microbridge weak links.

The equivalent circuit for the RSJ model is shown in figure 2.2 and the corresponding I–V characteristic in figure 2.3. In section 2.2 the analytical expression describing the I–V curve of a pure Josephson element shunted by a simple ohmic resistance was also given. The justification for using this model to describe a non-tunnelling weak link is mainly empirical, although GL theory has been used to provide a more rigorous justification. At first sight one might assume that a point-contact junction is just a narrow neck of metal possessing the same critical current density as the bulk superconductor. In reality the GL continuity relationships tell us that when the length of the junction is only of order $\xi(T)$, as is necessary if the Josephson effects are to be observed, dynamic processes will be occurring whenever a current flows, due to pairs being disrupted in the weak-link region on account of the high pair velocity gradient there. Since strong electromagnetic repulsion between current carriers ensures charge neutrality of any conductor down to a very short length scale, the total electron density at any point $n(r)$ will remain constant implying that the quasi-particle and pair densities, $n_1(r)$ and $n_2(r)$, are related by

$$n(r) = n_1(r) + n_2(r).$$

The Ginzburg–Landau equation (1.11) tells us how the amplitude of the order parameter ψ is related to the applied magnetic vector potential A and the current density in the weak-link region:

$$j = (e\hbar/2im)\,(\psi^* \nabla \psi - \psi \nabla \psi^*) - 2e^2\psi\psi^* A/m.$$

This expression can be solved to minimise the Gibbs free energy, showing that ψ is reduced as $\boldsymbol{j}$ increases, reaching zero at some critical value of the current density. The total current density $\boldsymbol{j}$ is the sum of the superconducting and normal components, these in turn being expressible in terms of pair and quasi-particle velocities $\boldsymbol{v}_1$ and $\boldsymbol{v}_2$, together with the number densities of the carriers:

$$\boldsymbol{j} = \boldsymbol{j}_1 + \boldsymbol{j}_2 = \boldsymbol{n}_1\boldsymbol{v}_1 + \boldsymbol{n}_2\boldsymbol{v}_2.$$

Since $n_1(r) = |\psi(r)^2|$ this quantity will also decrease with increasing pair velocity, reaching zero at the critical current density. The phase difference φ across the weak-link region is related to the integral of the pair velocity, thus:

$$\varphi = (2m/\hbar)\int \boldsymbol{v}_1 \cdot \mathrm{d}\boldsymbol{l}$$

(see equation (A.15)) so that n_1 is also a decreasing function of φ. When the critical supercurrent density is reached at the centre of the weak link ψ becomes momentarily zero and the phase change along the link becomes undefined. This constitutes a 'phase-slip' event in which φ changes abruptly by 2π, allowing the order parameter to re-establish itself at a finite value and the supercurrent to build up again. During this process the total current flowing remains constant but is switched from the pair channel to that of the normal current carriers, resulting in a transient voltage across the link accompanied by Joule heating. The time-averaged direct voltage is proportional to the rate at which 2π phase slips occur so that this 'relaxation oscillation' process gives a phenomenological explanation of the AC Josephson effects in a non-tunnelling junction.

Turning our attention now to the quasi-particle channel, the model assumes that these electrons behave ohmically, that is their contribution to the resistance is current-independent. Thus if the length of the link is l (with area A) and the mean scattering time is τ, j_2 is related to the direct voltage drop V along the link

$$j_2 = (n_2 e^2 \tau / ml)V.$$

Relating this model to the phenomenological resistance parameter R gives

$$1/R = n_2 e^2 \tau A / ml.$$

In addition to the variation of pair density with current there will of course be an intrinsic dependence of n_1 and n_2 on temperature. A simple two-fluid model (see section 1.3) suggests that

$$n_1 = n[1 - (T/T_c)^4]$$
$$n_2 = n(T/T_c)^4. \tag{2.32}$$

Substituting the latter expression into equation (2.31) suggests that $1/R$ should follow the same temperature dependence as n_2 since the scattering time should be approximately independent of temperature except for the very purest materials. Experimentally this is verified for microbridges provided they are short enough so that $l < \xi(T)$. However the situation is quite different for point contacts which exhibit a shunt resistance which is largely independent of temperature. This behaviour has not been explained adequately. Dissipation in both tunnel and point-contact junctions is a complex subject to which we will return in Chapter 8 when discussing macroscopic quantum phenomena.

2.6.4 Heating effects in Josephson junctions

There are a number of differences between point-contact and microbridge junctions, most of which have been explained in terms of variations in the thermal contact behaviour of the two types. In the microbridge case the main heat sink path is through the superconducting film itself. There are large thermal boundary resistances between film and substrate and between film and say liquid helium bath, even when the film is directly immersed in the cryogenic fluid. Since the film is essentially two-dimensional the heat flow along it is not great. As a result of these factors a number of thermal effects have been observed in the I–V characteristics of microbridge junctions. The most common is the presence of sloping 'step' discontinuities in the finite voltage portion of the curves, in the absence of applied microwave radiation or other electromagnetic interactions. These effects are most noticeable in relatively long bridges ($l > 10\ \mu$m) and are thought to be explicable in terms of 'localised phase-slip centres'. Several independent phase-slip centres may exist at different points along the length of a bridge simultaneously. As the voltage is increased the power dissipated in the bridge becomes greater and the heat transfer processes from the film become less and less able to cope with it. Consequently the order parameter in the bridge region of the film is reduced in amplitude as the temperature is raised. This continues until at a region r which is particularly poorly heat sunk (probably near the centre of the bridge) the critical temperature is exceeded and $\psi(r)$ goes to zero. A phase slip results. The I–V plots of long bridges show regions of steadily increasing slope as more and more centres are brought into the phase-slip process. It is found that frequently there is a small region of hysteretic behaviour associated with each centre as it becomes activated, not surprisingly in view of the thermal origin of the effect.

In the case of point-contact junctions, thermal phase-slip effects are only rarely seen. This is because this type of device is not simply a two-dimensional structure, since in going away from the centre of the link the narrow neck broadens out rapidly in two directions at right angles to one another, before the relatively thick post and anvil regions are reached. Such a geometry is very much more efficient than a planar film at conducting away heat generated in the weak-link region.

3

Superconducting Quantum Interference Devices

3.1 Superconducting Ring with a Weak Link

In Chapter 2 we made use of a hypothetical arrangement of a strongly coupled superconducting ring, interrupted by a region of weak superconductivity, to derive the essential features of the Josephson effects. In practice this same arrangement is actually employed as the basis of one of the most useful superconducting devices developed so far, the single-junction, or RF, SQUID. (As mentioned in the Introduction, SQUID is an acronym for Superconducting QUantum Interference Device.) Here the essential physics behind the operation of these important devices will be described, and in the following section the practical realisation of an RF SQUID with its associated readout electronics will be explained. Chapter 4 will treat in a similar way DC SQUIDs, which are related devices possessing two weak links in a single ring, whose mode of operation is rather different.

3.1.1 The internal flux as a function of applied flux

In Chapter 1 we saw that the free energy U of a superconducting ring is a periodic function of the flux Φ linking it (see figure 1.4). As the strength of coupling is gradually weakened between the order parameters on either side of a small cross section of the ring the heights of local free-energy maxima separating adjacent minima are decreased. When the critical current of the weakened region $i_1 \sim \Phi_0/L$ (where as before L is the ring self-inductance) the height of each local maximum becomes comparable with the energy difference between adjacent local minima. The free energy U (a function of both the external flux Φ_x and the included flux Φ_i) takes the

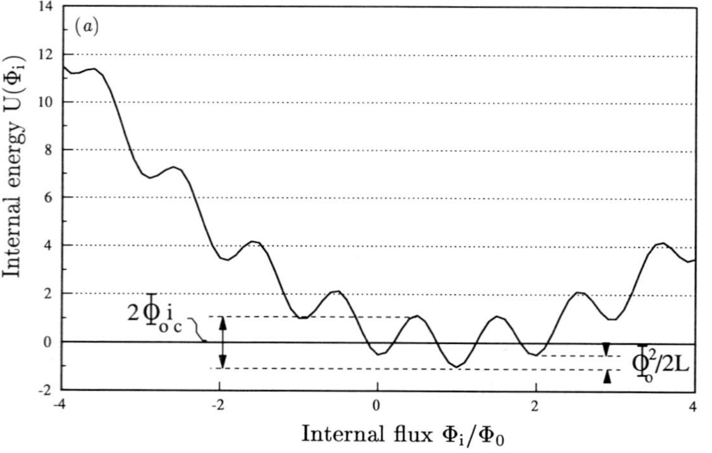

(a)
Internal energy U(Φi)
Internal flux Φi/Φ0
2Φoic
Φo²/2L
14
12
10
8
6
4
2
0
-2
-4
-2
0
2
4

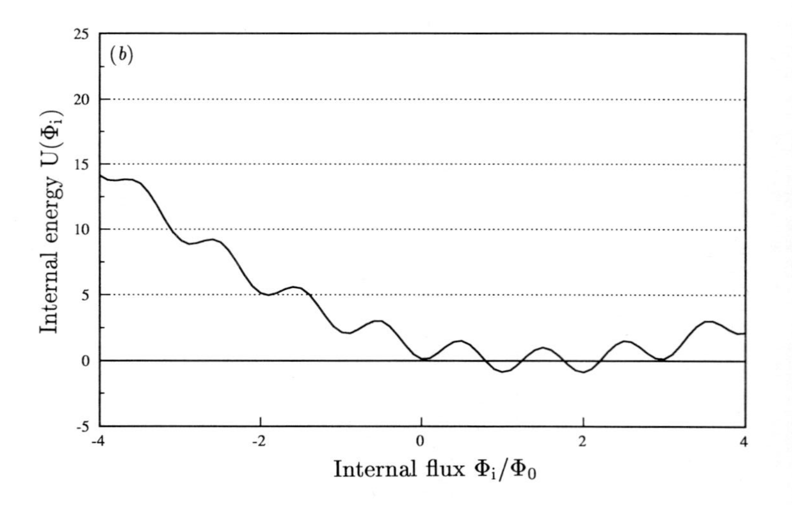

(b)
Internal energy U(Φi)
Internal flux Φi/Φ0
25
20
15
10
5
0
-5
-4
-2
0
2
4

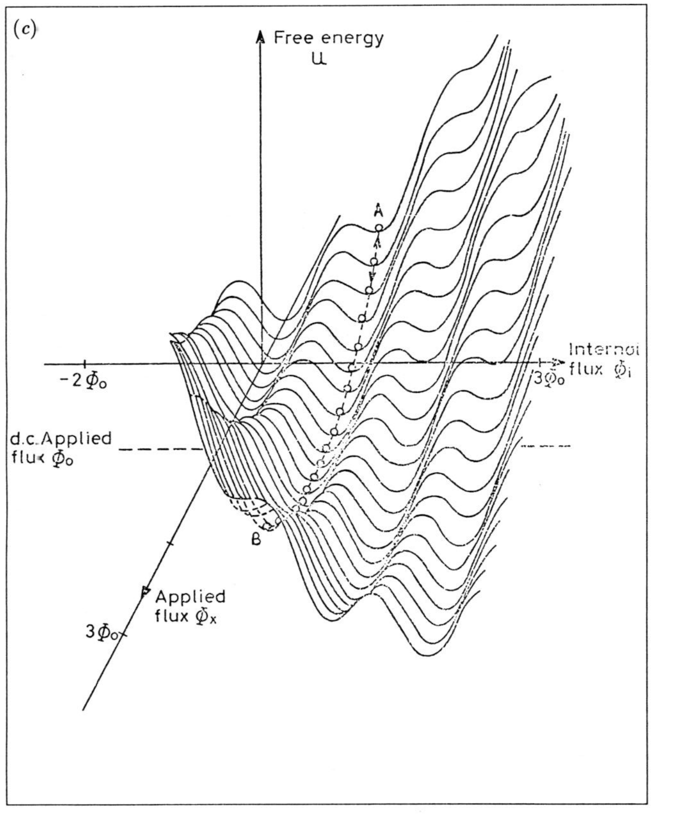

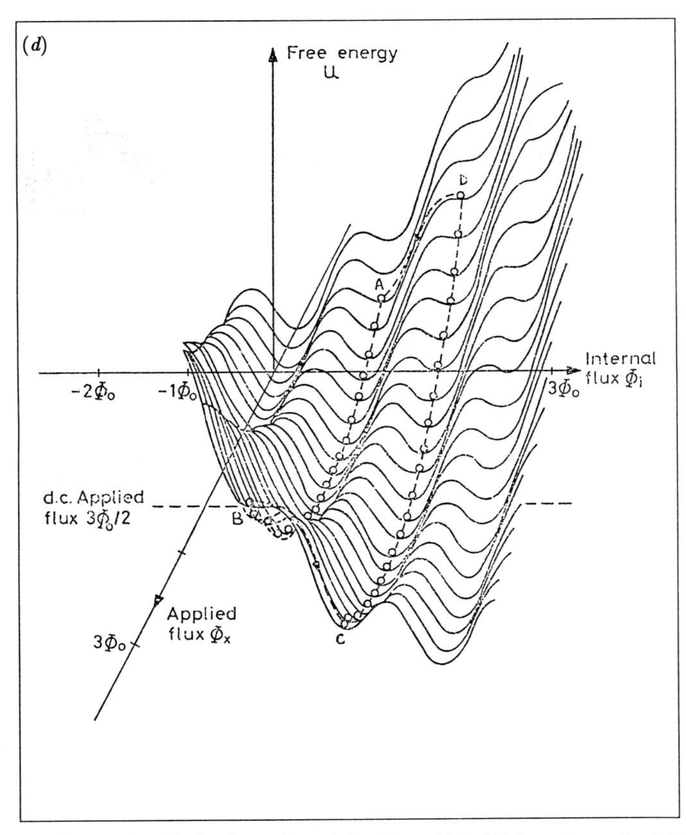

Figure 3.1 Internal energy $U(\Phi_i)$ for (*a*) $\Phi_x = 1\Phi_0$, and (*b*) $\Phi_x = \frac{3}{2}\Phi_0$. Applied flux for (*c*) $\Phi_0 + (3\Phi_0/2)\sin \omega t$, and (*d*) $3\Phi_0/2 + (3\Phi_0/2)\sin \omega t$

form

$$U(\Phi_i, \Phi_x) = -(\Phi_0 i_1/2\pi)\cos(2\pi\Phi_i/\Phi_0) + (\Phi_i - \Phi_x)^2/2L. \qquad (3.1)$$

The first term represents the coupling energy of the order parameter across the weak link and the second the stored energy in the ring resulting from the circulating supercurrent. The diagram 3.1(*a*) shows $U(\Phi_i)$ for a fixed value of applied flux $\Phi_x = 1\Phi_0$ whereas figure 3.1(*c*) shows the energy surface $U(\Phi_i, \Phi_x)$.

It has been traditional, though perhaps unjustified, to treat U and Φ_i as classical variables, so that at $T = 0$ the state of the SQUID may be represented by a particle rolling on the free-energy surface, constrained to that $U(\Phi_i)$ plane defined by the appropriate value of Φ_x. For the present we will assume the validity of the classical model but in Chapter 8 the far-reaching consequences of a quantum treatment of the ring variables will be described. Starting from the condition that $\Phi_x = 0$ with the particle at the absolute minimum of the free-energy surface, as the external flux is increased the particle will remain in that minimum, even when it becomes only a local minimum (see for example the path AB in figure 3.1(*c*)). For a classical particle at zero temperature there is no mechanism by which a transition to the new absolute minimum state can be brought about. According to this model the SQUID will not make a transition until Φ_x reaches a value such that $\delta U < 0$ for an infinitesimal change $\delta\Phi_i$ in Φ_i between successive points on the path going from the higher energy local minimum to the adjacent lower minimum, i.e. the intermediate energy barrier has reduced its height to zero. The value of the external magnetic flux required to reach this threshold condition depends more or less linearly on the ratio of the weak-link critical current to the inductance of the ring. (The critical current value determines the depth of the sinusoidal well, whereas the inductance sets the scale for the steepness of the parabolic background.) For this reason the dimensionless ratio $\beta = 2\pi L i_1/\Phi_0$ is important in characterising SQUID behaviour and it has been found that optimum detector performance is achieved for β values between about 1 and 10. The transition from one minimum to another as Φ_x is changed occurs when the circulating current i_s in the ring exceeds the critical current i_1 of the weak link. Figure 3.2(*b*) shows the sawtooth variation of i_s with Φ_x, for increasing and decreasing values. When the critical current is exceeded, a voltage pulse appears across the junction as the internal flux state of the ring changes by $\pm n\Phi_0$. The dynamics of flux quantum entry is, not surprisingly, very dependent on the normal-state resistance R of the weak-link junction. In fact R represents a macroscopic manifestation of the sum of all the diverse microscopic dissipative processes within the junction. It is these processes which absorb the excess energy associated with the transition from one flux state to a lower-energy one. Equivalently R also provides damping for the dynamics of this transition. If the junction is underdamped, once the transition is underway the internal flux may change by more than a single flux quantum ($|\Delta\Phi_i| \gg \Phi_0$).

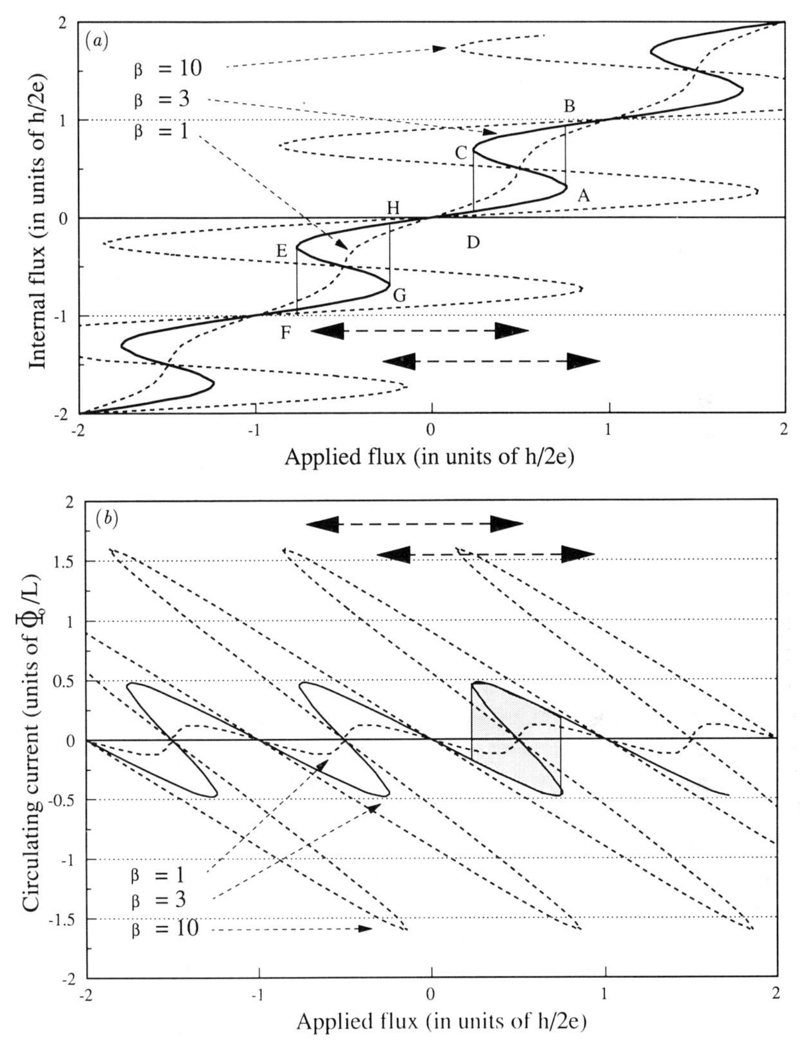

Figure 3.2 Plot of (*a*) the internal magnetic flux, Φ_i, and (*b*) the circulating current versus applied magnetic flux Φ_x

3.1.2 Operation of RF SQUIDs

We have considered the free-energy surface of a superconducting ring interrupted by a weak link, and the dynamics of flux state transitions. To extend from this model to the basis of operation of the single-junction SQUID now involves the application to the ring of an oscillating external flux of such an amplitude that for the time-averaged value of $\langle \Phi_x \rangle$ (shown in figures 3.1(*c*) and 3.1(*d*)) the SQUID is moved between the limits A and B marked on the free-energy surface diagram. Provided $\langle \Phi_x \rangle \sim n\Phi_0$ (n an integer), the amplitude of the AC flux applied is insufficient to cause a transition from one internal flux state to another. However if $\langle \Phi_x \rangle \sim (n+\frac{1}{2})\Phi_0$ then the same amplitude of alternating flux *is* now sufficient to cause the SQUID to make internal flux state transitions. Consider now an overdamped SQUID so that $|\Delta\Phi_i| = \Phi_0$. We see from figure 3.1(*d*), which shows the $U(\Phi_x, \Phi_i)$ surface for $\langle \Phi_x \rangle = 3\Phi_0/2$, that the path taken by the SQUID is now ABCD. The paths B to C and D to A correspond to one flux quantum entering and leaving the ring respectively. Thus a small change of only $\Phi_0/2$ in Φ_x is capable of bringing about these transitions of internal flux state. Given the small size of $\Phi_0 = 2 \times 10^{-15}$ Wb, all that is required to produce a sensitive flux detector is a means of detecting the onset of these transitions as $\langle \Phi_x \rangle$ is slightly altered.

The key result from section 3.1 may be summarised as follows: a small change in the direct flux applied to a single-junction SQUID ring is capable of causing internal flux state transitions of the SQUID ring in the presence of an applied AC flux modulation at a frequency ω, of amplitude $\sim 2\Phi_0$ peak to peak.

3.2 RF SQUID Detection Systems

In this section we consider how the change in the rate of internal flux state transitions might be detected. Of course the change cannot be sensed directly by a conventional flux detector, if only because the main purpose of building a SQUID is to produce a sensor capable of higher flux sensitivity than any other device. Suppose instead a detector was coupled to the SQUID ring which could sense the increased energy dissipation arising from transitions from one internal flux state to an adjacent one. The amount of energy dissipated at each transition is of order $\Phi_0^2/2L$ (equation (3.1)), so for two transitions per cycle of the RF drive the power dissipated in the SQUID ring will be of order $\omega\Phi_0^2/2\pi L$. Inserting reasonable values ($\omega/2\pi = 10^9$ Hz, $L = 0.5$ nH) yields a power of around 10 pW. A change of much less than $\Phi_0/2$ in the DC applied flux is capable of producing this change in power dissipation, which should be easily detectable using some form of cryogenic bolometer. In practice, as far as the author is aware this method of detection has never been tried, although it might shed some light on the nature of dissipation in SQUID rings (see section 8.6). Instead, a method of resonant detection, used in the first demonstration of a single-junction SQUID (Silver and Zimmerman 1967), has become almost universally adopted. This is somewhat analogous to the detection methods used in radio receivers and NMR systems.

3.2.1 Resonant detection system

The experimental simplicity of this method probably accounts for its popularity. At its simplest the system consists of a lumped component parallel resonant LC circuit (the 'tank circuit'). The inductor is in the form of a small air-cored solenoid which is inserted through the superconducting ring. A low-loss capacitor is connected in parallel with this coil, its value being chosen so that the circuit resonates at some convenient radio frequency, typically 20–30 MHz. An oscillatory flux modulation can then be applied to the SQUID ring via its mutual inductance M with the coil of the resonant circuit. If this parallel circuit combination is fed from a current source at the resonant frequency ω_0, in the steady state an alternating current will flow in the coil, thereby producing an alternating flux in the SQUID ring of amplitude $MQi \sin \omega_0 t$, where Q is the quality factor of the resonant circuit and $i \sin \omega_0 t$ is the current fed from an external current source to the resonant circuit. The DC flux $\langle \Phi_x \rangle$ to be sensed is applied via an additional coil, also inserted through the ring.

To measure a change in $\langle \Phi_x \rangle$ it is now only necessary to amplify and detect any change in the RF voltage appearing across the resonant circuit. Suppose that initially $\langle \Phi_x \rangle \sim n\Phi_0$ and i, the RF current amplitude, is set just too low to produce internal flux state transitions. As a result power dissipation in the SQUID will be at a minimum, leading to maximum RF voltage across the circuit, that is the Q (quality factor) of the circuit is set almost entirely by dissipation in the inductor and capacitor. When $\langle \Phi_x \rangle$ is changed to $(n + \frac{1}{2})\Phi_0$ the SQUID begins to undergo frequent changes in its internal flux state. The energy dissipated by these transitions is drawn from the energy stored in the tank circuit. The result is a drop in the mean RF voltage amplitude V_{RF}, or equivalently a reduction in Q. A simple system of low-noise high-gain RF amplification, followed by detection, gives a direct voltage which is periodic in $\langle \Phi_x \rangle$.

This rather naive description of the operation of the resonant detecting method for single-junction (or RF) SQUIDs leaves many unanswered questions, to which we return later. However it is clear that the voltage response will be periodic in the DC applied flux, but we need to know the answers to such questions as how the detected RF voltage will vary between the two extreme values discussed for integer or odd half-integer flux quanta values. How is the RF drive current amplitude adjusted to the correct level? What influence does the narrow-band frequency response of the parallel resonant circuit have on the general operation of the SQUID? What are the noise levels in the SQUID itself and how good must the following RF amplifier be if its noise is not to degrade the overall sensitivity? These questions will be considered in section 3.3, which the reader may prefer to omit on a first reading.

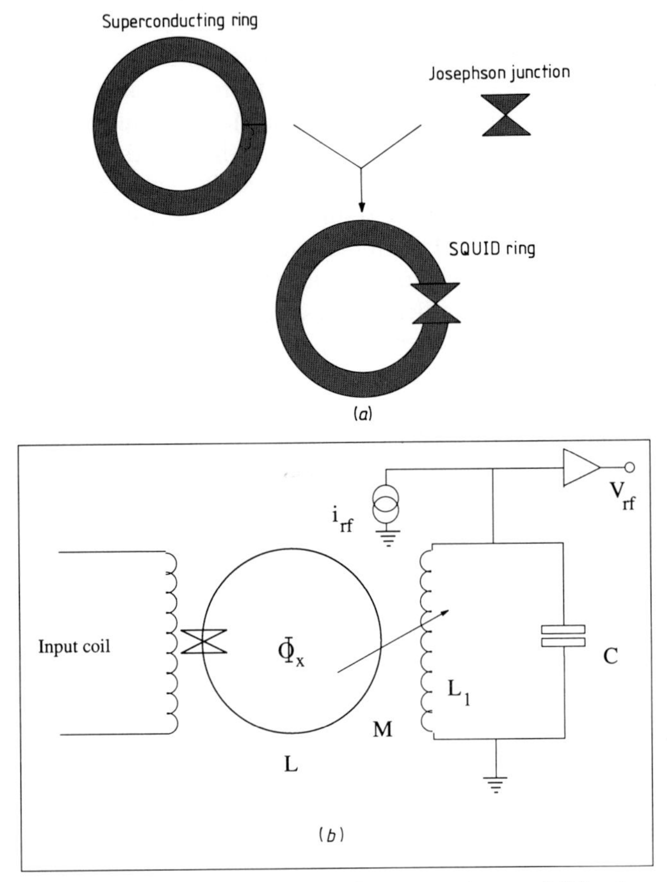

Figure 3.3 (*a*) Basic components of a SQUID ring, and (*b*) schematic circuit diagram of an RF SQUID

A schematic circuit diagram of an RF SQUID is shown in figure 3.3, and the general form of the variation of V_{RF} with $\langle \Phi_x \rangle$ for different RF current drive levels is shown in figure 3.4. Note the essentially triangular response of the voltage. This smooth variation of $V_{RF}(\langle \Phi_x \rangle)$ means that as it stands the RF SQUID is a linear detector of magnetic flux, provided that the changes in flux are limited to $\pm \frac{1}{2}\Phi_0$. To produce a linear device with a greater range, negative feedback can be used. An audio-frequency magnetic flux modulation of amplitude $\pm \Phi_0/2$ is applied to the ring—usually via the resonant circuit coil. The detected RF voltage is fed to a lock-in amplifier, referenced by the audio modulation signal. The output from this device can then be used to apply feedback flux, via a feedback resistor and the RF coil, and provided the gain of the feedback loop is sufficiently high the ring will now operate with a fixed DC flux (the 'flux-locked' mode). In this condition the output from the lock-in will be linearly related to the change in the signal flux, even when this is much greater than a flux quantum. A circuit diagram of the flux-locked system is shown in figure 3.5. As an indication of the performance to be expected from a SQUID detector, that of a commercially available system using a frequency of around 20 MHz might be taken as typical. RF components are readily available for this frequency and their performance has been optimised so the implementation of high-quality electronics is straightforward. Such a SQUID is usually limited by noise in the preamplifier, but flux resolution as small as $5 \times 10^{-5} \Phi_0 \, \mathrm{Hz}^{-1/2}$ in a measurement bandwidth of 30 kHz is typical.

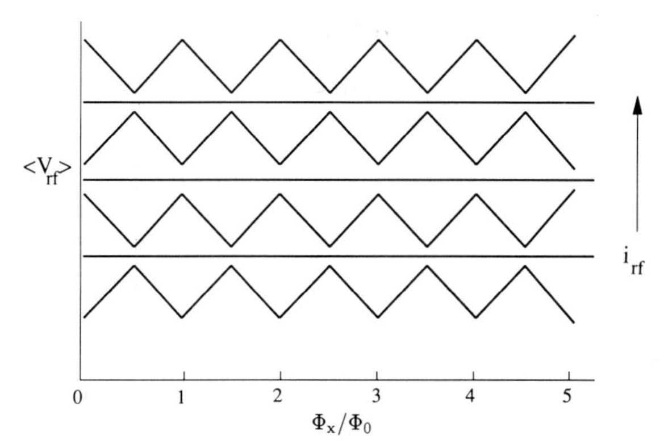

Figure 3.4 Detected RF voltage as a function of applied flux for various increasing RF current levels

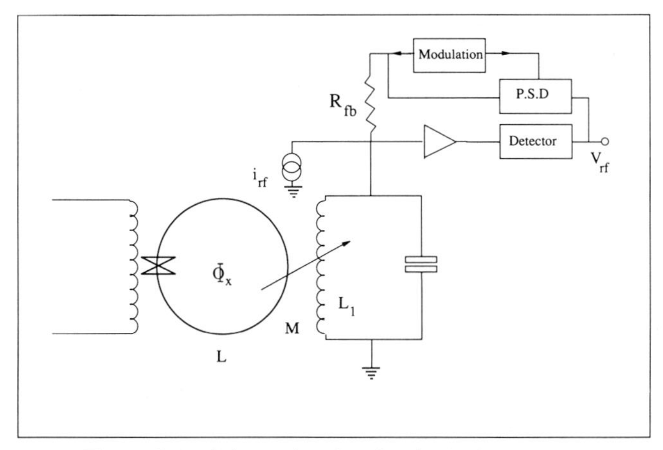

Figure 3.5 Schematic of a flux-locked RF SQUID

3.3 Quantitative Treatment of RF SQUIDs

We have taken a somewhat unusual approach in the first two sections of this chapter, by describing the basic mode of operation of the single-junction SQUID in terms of its free energy $U(\Phi_x, \Phi_i)$. However this is not the only model capable of explaining the functioning of these devices. In fact it has been customary to describe RF SQUIDs purely in terms of the way in which Φ_i varies with Φ_x. Such a treatment is capable of giving a quantitative description of the SQUID parameters and will now be briefly outlined. In the author's view the free-energy picture is actually the more fundamental model. Its use is justified not only on these grounds, but also that it provides an appropriate introduction to a macroscopic quantum mechanical description of a SQUID, which will be introduced in Chapter 8.

In section 2.2 a relationship between the circulating supercurrent in the ring i_s and the internal flux Φ_i was derived (equation (2.8)):

$$i_s = i_1 \sin \varphi = i_1 \sin[2\pi(1 - \Phi_i/\Phi_0)] .$$

But also the external flux Φ_x is linked to these two parameters by an additional relationship

$$\Phi_i = \Phi_x + Li_s. \tag{3.2}$$

Thus we have two equations involving $i_s(\Phi_x)$ and $\Phi_i(\Phi_x)$. The non-linear term in equation (2.8) means that no general analytic solution to these can be given. However graphical or numerical solutions are straightforward. Figure 3.2 shows the dependence of $\Phi_i(\Phi_x)$ and $i_s(\Phi_x)$ for three different values of the parameter β. The plots indicate that for $\beta < 1$ the flux and circulating current are single-valued functions of Φ_x, but when $\beta > 1$ multi-valued behaviour is shown as the graphs become re-entrant. This re-entrant behaviour indicates that metastable states exist in the free-energy model, a necessary condition for the resonant detection scheme outlined above. Consider the effect on the internal flux state of an applied external flux modulation. If $\langle\Phi_x\rangle \sim n\Phi_0$ the modulation amplitude indicated in figure 3.2 by the double-headed arrow is insufficient to bring about a change in Φ_i. If however $\langle\Phi_x\rangle \sim (n + \frac{1}{2})\Phi_0$, as before, then the same amplitude of modulation will cause internal flux transitions. In this case the path ABCD will be traversed thereby showing that Φ_i has a hysteretic dependence on Φ_x. The energy absorbed in executing one such hysteresis loop is

$$dU = -\int i_s \, d\Phi_x.$$

This integral is represented by the sum of the two shaded areas in figure 3.2(*b*), and is clearly of order $\Phi_0{}^2/2L$, as stated in the previous section. It is customary to assume the energy is dissipated as heat in the Josephson junction, although there seems to be no direct evidence of this. If the ring is driven by a resonant circuit then taking the SQUID around the flux hysteresis loop will draw this amount of energy from the inductor, reducing the level of oscillation of the tank circuit and the resulting amplitude of Φ_x applied to the SQUID. It will take a number of cycles of the RF field for the current in the inductor to build up again to the level at which it is great enough to cause further loops to be executed. In fact the definition of the quality factor Q of the circuit tells us that approximately Q cycles of the oscillation will be required to restore the level of oscillation. When the threshold level is again reached the ring will execute another hysteresis loop, extracting further energy which then has to be built up again. This 'relaxation oscillation' process typifies the resonant dissipative mode of operation of the SQUID.

3.3.1 *The linear approximation*

The absence of analytical solutions for $\Phi_i(\Phi_x)$ makes it tedious to derive quantitative results from this model. Instead we may make a reasonably realistic linear approximation, namely that

$$\Phi_i = \Phi_x + Li_s = n\Phi_0 \qquad (i_s \leqslant i_1). \tag{3.3}$$

This is equivalent to neglecting the phase shift φ across the weak link, except when a transition occurs. Consider the LC circuit driven by an RF current source at ω_0, the circuit's resonant frequency. If the mutual inductance between the ring and inductor is M and the current flowing in the coil is i, $\Phi_x(t) = Mi(t)$, then as long as the internal flux state remains the same there will be a simple linear relationship between i and i_s:

$$i_s = -Mi/L. \tag{3.4}$$

While n remains fixed because the amplitude of i is too small to cause transitions, there will be just two inductive contributions to the voltage V appearing across the inductor:

$$V = \mathrm{i}\omega L_1 i + \mathrm{i}\omega M i_s = \mathrm{i}\omega L_1 i(1 - k^2) \tag{3.5}$$

where k is the coupling coefficient between ring and coil, given by

$$M^2 = k^2 L_1 L. \tag{3.6}$$

In practice the ring is weakly coupled to the coil so that $k^2 \ll 1$ (in practice one should aim to satisfy the condition $k^2Q \sim 1$ for reasons which will be explained below). The initial build-up of RF current in the inductor is set by the expression

$$i = Q[1 - \exp(-\omega_0 t/2Q)]\, i_0 \sin \omega_0 t. \tag{3.7}$$

As soon as this current exceeds the threshold value $i_1 L/M$ then $i_s > i_1$ and the SQUID ring makes a transition to a neighbouring internal flux state $(n+1)$, in a time of order L/R (the time constant of the ring for normal current flow). For a typical SQUID weak link $R > 1\ \Omega$ and $L \sim 1$ nH so that $L/R < 1$ ns and this time is much shorter than one period of the RF oscillation so the transition takes place at effectively constant flux through the resonant circuit inductor. At any instant the total flux through this component is

$$\Phi_t = iL_1 - iM^2/L + i_s M.$$

When the SQUID ring makes a transition of $1\Phi_0$ the change in i_s is $|\delta i_s| = \Phi_0/L$. The resulting change in the inductor current is

$$|\delta i| = -\Phi_0 k^2/(M(1-k^2)). \tag{3.8}$$

Half a cycle later the ring undergoes another flux change in the opposite sense, since the amplitude of i required for this reverse transition is less than for the forward one. Energy will again be drawn from the tank circuit, reducing the amplitude of i as before, the current then growing again as indicated by equation (3.7). Thus i builds up at a rate which depends on i_0. However the average level of oscillation is more or less independent of i_0, for as i_0 increases the threshold level is reached more quickly so that the value of i is reset more frequently to its base value. Figure 3.6 shows this build-up schematically for two values of i_0. The average voltage across the circuit inductor under this condition is

$$\langle V \rangle = \omega_0 L_1 (1-k^2) L i_1 / M = \omega L_1 Q i_0 / (2^{1/2}). \tag{3.9}$$

The maximum rate of energy absorption occurs when, on each cycle of the RF excitation, a hysteresis loop is executed. Since no further increase in dissipation is possible, any additional increase in excitation level will now cause the mean voltage to increase again. Until, that is, the level is high enough that an additional double hysteresis loop (shown in figure 3.2(*a*) by the overall path EFGHDABC) can be executed. The same argument may be applied again to show that another flat step in the plot of $\langle V \rangle$ against i_0 will follow. Increasing the excitation level i_0 further produces still more steps as higher multiple loops are brought in. The full curve of figure 3.7 shows the mean RF voltage $\langle V \rangle$ as a function of the mean RF current amplitude $i_0/(2^{1/2})$ applied to the resonant circuit.

So far we have assumed in this argument that the applied flux consists of only an alternating component. If in addition there is a time-independent contribution $\langle \Phi_x \rangle$ to $\Phi_x(t)$ the onset of internal flux state transitions will begin when the sum of the alternating and constant terms is just equal to Li_1. Using equations (3.4) and (3.5) this is seen to occur when the amplitude of the RF voltage becomes

$$|V| = (\omega_0 L_1/M)\ (1-k^2)\ (Li_1 - \langle \Phi_x \rangle). \tag{3.10}$$

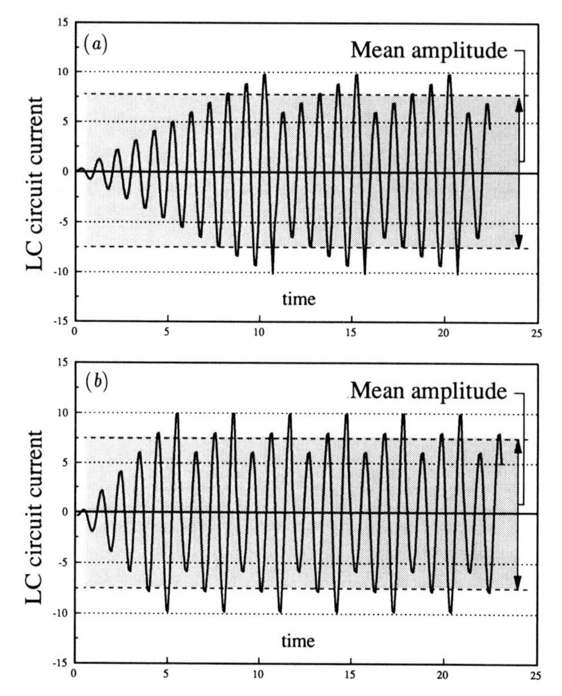

Figure 3.6 Build-up of *LC* circuit current with time, after switch-on at $t = 0$. (*a*) I_{RF} for SQUID on first step, and (*b*) I_{RF} greater, still on same step, more rapid recovery but same average amplitude in steady state

As before, a step occurs in the *V–I* characteristic at this voltage and, as before, higher RF current excitation levels will produce steps at increasing voltages. The step voltage corresponding to a particular flux transition is at a *maximum* whenever the applied DC flux is equal to a *whole number* of flux quanta, and at a *minimum* for *odd half integral* values. Note the first step for odd half-integer values of Φ_x is only half the length of subsequent steps since only one hysteresis loop per cycle is executed. Subsequent odd-numbered steps correspond to $3, 5, 7, \ldots$ loops. To operate the device the RF current is adjusted so that the SQUID is biased onto a constant voltage step for all values of DC flux (figure 3.7 shows that this is possible in typical cases). The voltage modulation amplitude δV is the range of variation in $\langle V_{RF} \rangle$ which is produced as Φ_x is changed and is given by

$$\delta V = \omega_0 \Phi_0 L_1 (1 - k^2)/2M. \tag{3.11}$$

It appears from this expression that the voltage modulation amplitude can be increased without limit by decreasing M. In fact this is not the case. If the coupling is made too weak the Q of the circuit will be scarcely affected at all by the energy dissipated in the SQUID ring. Equation (3.11) does not bring out the point that the coupling constant k and the quality factor Q are interdependent. We can simply show that the maximum voltage modulation occurs if the relation $k^2Q = 1$ is approximately satisfied. This is the explanation of why the coupling between RF coil and SQUID ring should be weak since the natural Q of an air-cored RF coil will be of order 50–100.

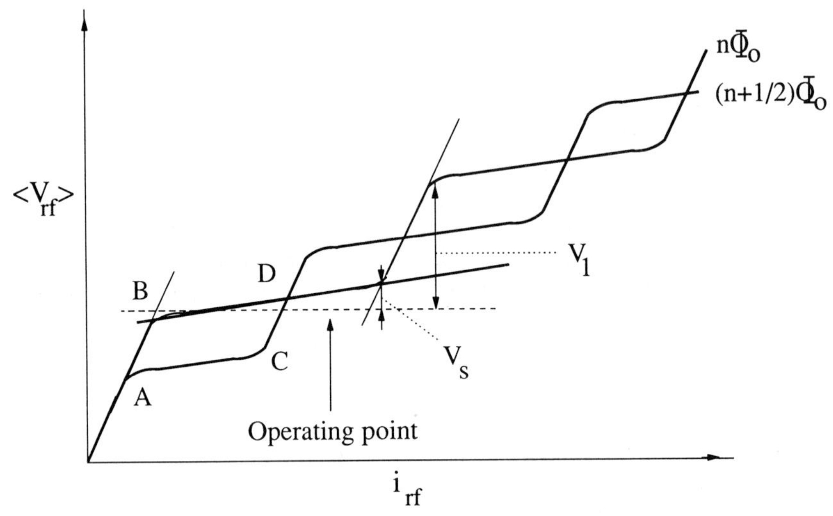

Figure 3.7 *I–V* characteristic of RF SQUID showing the definition of the fractional step-rise parameter V_s

Consider the voltage–current characteristic shown in figure 3.7. The slope of the characteristic between constant voltage steps (e.g. the sections such as AB, CD etc) is given by

$$V/i_0 = \omega_0 L_1 Q(2^{1/2}) \tag{3.12}$$

whereas the length of each full constant voltage step is

$$\delta i_0 = 2\Phi_0 k^2/M(1 - k^2). \tag{3.13}$$

Comparing these two equations we see that when $k^2Q = 1$ the point B, representing the start of the step for integer quanta, lies above the point C, the end of the step for half-integer quanta. Any further reduction in k, and hence M, will move the point B even further to the right. The full voltage amplitude corresponding to the step separation will then no longer be attainable. Thus the condition $k^2Q = 1$ is the optimum one which maximises the flux to voltage transfer ratio $(\partial V/\partial\Phi)$ for the RF SQUID. We may simply show that the power gain of the RF SQUID, regarded as a linear amplifier, is proportional to the pump frequency ω_0. For a change $\delta\Phi$ of the DC flux at the input the available input energy is $dE = (\delta\Phi^2/2L)$ and if this change takes place in a bandwidth df the available signal power is just $dE \times df$. The available output power resulting from such a change is

$$\delta P = (\partial V/\partial\Phi)^2(\delta\Phi^2/2R) \tag{3.14}$$

where the voltage to flux transfer ratio is given by equation (3.11). So the power gain g, that is the ratio of output to input power, simplifies to

$$g = \omega_0 Q^2/df \tag{3.15}$$

(we have used the relationship $Q = \omega_0 L/R$ for a parallel resonant circuit.) The expression for g is reminiscent of the gain factor of a parametric amplifier (embodied in the Manley–Rowe relationships (Manley and Rowe 1956)), and it has been shown in detail that an RF SQUID is an unusual type of upconverter (Ehnholm 1977). Unlike traditional parametric amplifiers which conventionally employ a capacitor, the parametric element of the SQUID is its Josephson inductance. For a radio-frequency SQUID operated at 30 MHz with a Q of 50 and a signal bandwidth of 1 kHz the power gain becomes 5×10^8, a remarkably high figure for any single-stage linear amplifier.

On the simplest assumption that the intrinsic noise limiting its performance might be expected to be frequency-independent we would be led to suppose that the effective sensitivity of an RF SQUID would increase with pump frequency. Further discussion of the limiting sensitivity of these devices will be deferred to the last section of this chapter. However it is a matter of fact that most RF devices have operated with a pump frequency of around 10–30 MHz, for which low-noise amplifiers are readily available. Higher-frequency amplifiers covering the UHF range are now available at a low price so that recently some high-performance SQUIDs have been described, at least one of which uses a liquid-helium-cooled GaAs FET preamplifier (Prance *et al* 1981). SQUIDs pumped at microwave frequencies have also been used, although the reported performance is not as good as that at UHF. This reflects the deteriorating noise figures of amplifiers as the frequency range rises into the microwave region. Additionally the high-frequency properties of the Josephson element have not been considered in the simplified model used above, and it will require very well characterised circuit elements, including the Josephson junction, to produce an improvement in sensitivity proportional to the frequency increase beyond the UHF region.

3.3.2 Noise in RF SQUIDs

In the following section we attempt to quantify practical limits to the detection sensitivity of SQUIDs. The noise processes which limit the measuring sensitivity of these devices are numerous and complex; some are intrinsic to the SQUID ring while others originate in the circuitry associated with it.

3.3.3 Thermal noise in the SQUID ring

For a superconducting ring carrying a direct circulating supercurrent which is much less than its critical value, the average energy of a thermal fluctuation ($kT/2$ at temperature T) is insufficient to excite the ring to an internal flux state of higher energy. According to a classical model, which describes the SQUID free energy by a sinusoidally periodic structure, superimposed on a parabolic potential curve (see figure 3.1), thermal fluctuation energy will only allow the state to 'wobble' around in the metastable local minimum. The operation of an RF SQUID depends on the flux sweep reducing the barrier between adjacent flux states until, at some point on the modulation cycle, it disappears. At 0 K in a classical picture it is at this point that the flux state transition occurs. For $T > 0$ thermal fluctuation energy will allow the transition to occur earlier than expected in the flux sweep cycle. This in turn affects the amount of energy drawn from the LC circuit (given by the area of the Φ_i versus Φ_x loop (figure 3.6)), and $\langle V_{RF} \rangle$ will likewise be reduced. Such fluctuations will appear as random noise at the SQUID output and will be indistinguishable from flux noise originating at the input coil. The non-linear character of the device ensures that the simple white spectral distribution of thermally generated noise fluctuations will be significantly modified. Detailed numerical calculations of the equivalent flux noise due to this process have been carried out (Kurkijarvi 1972) and have been confirmed using a 19 MHz point-contact SQUID down to 2 K (Jackel *et al* 1972, 1974). Figure 3.8 shows data for the probability distribution for flux state transitions as a function of DC applied flux, taken from the same reference. Note that the width of the distribution is large, of order $0.1\Phi_0$, although much lower than that under normal RF bias conditions. An approximate analytical form for the spectral density $S(\omega)$ of equivalent flux fluctuation has been given:

$$S(\omega) \sim \Phi_0{}^2(\pi/\omega_0) \times [(2\pi kT/\Phi_0 i_1)\beta(1-\beta^{-2})^{1/6}/2\pi]^2. \qquad (3.16)$$

The same thermal excitation process is responsible for another feature of the RF SQUID I–V characteristic. When the SQUID is biased onto a step region (say the section BC in figure 3.7) $\langle V \rangle$ will be higher towards the upper end of the step than at its onset. This is because the flux sweep rate is higher here so that the probability of an early transition is reduced. The result is that the step possesses a slope $\alpha = V_1/V_s$ (see figure 3.7) which is temperature-dependent and is given by the approximate expression

$$\alpha \sim (\omega_0 S(\omega)/\pi\Phi_0{}^2)^{1/2}. \qquad (3.17)$$

Measurement of α provides a simple method for determining the intrinsic SQUID equivalent flux noise spectral density $S(\omega)$. The parametric description of the RF SQUID (Ehnholm 1977) mentioned above has given useful insight into some subtle sources of noise. When feedback mode is used small changes in input flux ($\delta\Phi_1$) and output voltage (δV_2) are related to changes in input flux (δi_1), (δi_2) by the matrix equation:

$$\begin{pmatrix} \delta\Phi_1 \\ \delta V_2 \end{pmatrix} = \begin{pmatrix} \partial\Phi_1/\partial i_1 & \partial\Phi_1/\partial i_2 \\ \partial V_2/\partial i_1 & \partial V_2/\partial i_2 \end{pmatrix} \begin{pmatrix} \delta i_1 \\ \delta i_2 \end{pmatrix}. \tag{3.18}$$

The various transfer functions in the 2×2 matrix may be calculated from SQUID parameters. Of particular interest is the term $\partial\Phi_1/\partial i_2 = \pi L_2 M_1/4M_2$ where the parameters are as indicated in the equivalent circuit of figure 3.5. It represents the reverse transfer function which indicates how noise in the output circuit will appear as effective noise at the input (this is the so-called 'back-reaction' noise which is of considerable importance in gravity wave detectors, see Chapter 8).

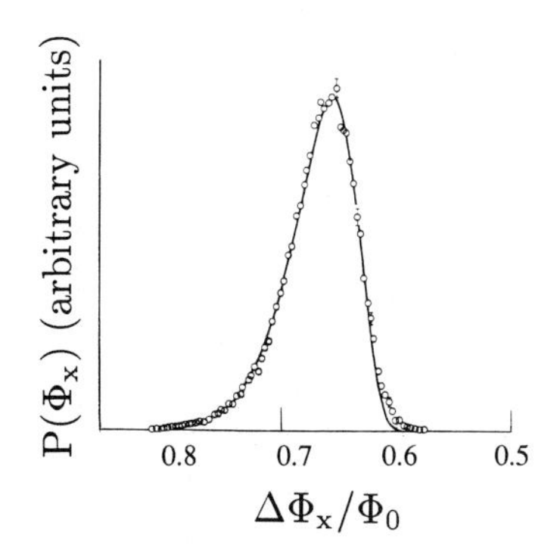

These are thermally activated. The probability of a transition within a given interval of Φ_x is plotted as a function of the difference in Φ_x between where the transition actually occurs and where it would occur at absolute zero.

(After Jackel et al. (1972))

Figure 3.8 Experimental distribution of SQUID flux state transitions

3.3.4 Tank circuit and amplifier noise in RF SQUIDs

The quality factor Q of the RF resonant circuit must not be so great that it can severely limit both the frequency response and the dynamic range of the SQUID. We have seen that the condition $Qk^2 \sim 1$ should be approximately satisfied. The loss mechanisms which produce a finite Q value also cause thermal-fluctuation-generated noise voltages to appear across the tank circuit, which are independent of thermal noise originating in the SQUID ring. Similarly the sensitive preamplifier following the resonant circuit will have a finite noise temperature. For optimum power transfer the real part R of the input impedance of the preamplifier and the dynamic impedance of the loaded resonant circuit $\partial V_2/\partial i_2$ should be approximately equal. In this case the available signal power δP for an input flux change

$\delta\Phi_1$ becomes

$$\delta P = 2\omega_0(\delta\Phi_1)^2/\pi L_1. \tag{3.19}$$

If the noise temperature of the preamplifier is T_p the minimum power detectable in a bandwidth of 1 Hz is just kT_p, so that the equivalent flux noise per unit bandwidth, referred to the input, is:

$$\delta\Phi_1 = (\pi L_1 kT_p/2\omega_0)^{1/2}. \tag{3.20}$$

In general for most practical devices the ratio r of the spectral density of flux noise due to the amplifier and that due to intrinsic processes is just

$$r \sim 2L_1 kT_p/2\alpha^2\Phi_0^2 \gg 1$$

so that intrinsic processes are almost never dominant. For example, for radio-frequency-pumped devices the amplifier noise temperature must be <30 K before intrinsic effects become important whereas the best room temperature amplifiers have noise temperatures approximately three times higher.

3.3.5 Flux-locking feedback circuits

As has already been pointed out in section 3.2 the flux periodic response of the basic SQUID ring does not make for a very useful general purpose detector. By employing negative feedback the response may be linearised. For a flux modulation of amplitude $\pm\Phi_0/2$ applied at an audio frequency ω_2 the output from the tuned circuit, following amplification and detection, is fed to a phase-sensitive detector (PSD), referenced by the modulation frequency. Then after further signal conditioning which, at its simplest, consists of a simple integrator with a cut-off frequency ω_c, the signal is inverted and fed back to the tank circuit coil through a large resistor R_2. If the forward transfer function relating the output V_2 following the integrator to the signal flux $\Phi_1(\omega)$ is $h = \partial V_2/\partial\Phi_1$ then

$$V_2 = \Phi_1 R_2/(1 + \mathrm{i}\omega t) \tag{3.21}$$

where t^{-1} is the unity gain frequency of the loop, given by

$$t = R_2/Mh\omega_c.$$

For sufficiently large values of $h(\omega)$ the feedback signal will maintain the total low-frequency flux constant at a value close to a half-integral number of quanta. In this fixed flux state the SQUID is said to be 'flux-locked'. As is usual for simple feedback circuits, the loop gain must be reduced to less than unity at the frequency for which the phase shift around the loop reaches the value π, otherwise oscillation will occur. Stability, in the case of a flux-locked SQUID circuit, is harder to achieve since any of the states which corresponds to a half-integer number of quasi-static flux quanta through the ring is an allowed stable state. Extraneous noise must be reduced to such a level that it is not capable of inducing such changes of state. The maximum signal amplitude which can be tolerated at the modulation frequency is only $\Phi_0/4t$. The parameters of the loop also determine the maximum speed with which the input flux may be changed (the slewing rate). This is defined as the maximum rate at which the feedback flux can change, which from the above expression for V_2 is readily seen to be

$$(\mathrm{d}\Phi_1/\mathrm{d}t)_{\mathrm{max}} = \Phi_0/4t.$$

The same argument shows that the maximum signal amplitude which may be detected without losing the flux-locked condition at any frequency is $h(\omega)\Phi_0/4$. Commercially manufactured SQUIDs use more sophisticated feedback circuits than the simplest arrangement described here, in order to optimise the slewing rate while still ensuring stable performance, but these are beyond the scope of this treatment.

3.4 Practical RF SQUIDs

Although the first SQUIDs were thin film devices the problem of coupling flux effectively into a planar structure, which requires a multi-turn coil, soon led to the development of bulk structures, employing simply made, although not very stable or predictable, point-contact junctions.

3.4.1 The two-hole SQUID

The design which was favoured for some years was the 'two-hole' SQUID due to Zimmerman *et al* (1970) (figure 3.9), which is essentially two superconducting rings joined by a common Josephson junction. The body is made from a cylinder of niobium, about 20 mm long and 8 mm diameter, through which two axial holes (~3 mm ∅) are drilled. One hole contains the signal coil to which external sources may be coupled, while the other contains the RF tank circuit inductor, a tuning capacitor being connected externally across a coaxial line connecting to the room temperature preamplifier. A thin slot joins the two holes and is bridged by a pointed adjustable Nb screw which lightly contacts onto an Nb anvil, forming the Josephson junction. Following adjustment at room temperature, lock nuts secure both point and anvil. One strong point of the design is that all parts are made from a single material (usually niobium) so that differential thermal contraction is not a problem if cooling is carried out slowly enough to ensure reasonable temperature uniformity across the device. In view of the fragile nature of the point contact, thermal and temporal stability of these SQUIDs is remarkably good.

The two-hole symmetric SQUID has another significant advantage over earlier designs in that it is 'self-screening'. That is to say the entire arrangement can be seen as two superconducting loops in parallel, each interrupted by a weak link, but both being surrounded by a complete superconducting ring in the form of a long niobium tube. This tube very effectively attenuates external magnetic field fluctuations so that with no input coil connected the device may operate without any additional superconducting shield. It was the success of this construction which largely led to the design being produced commercially.

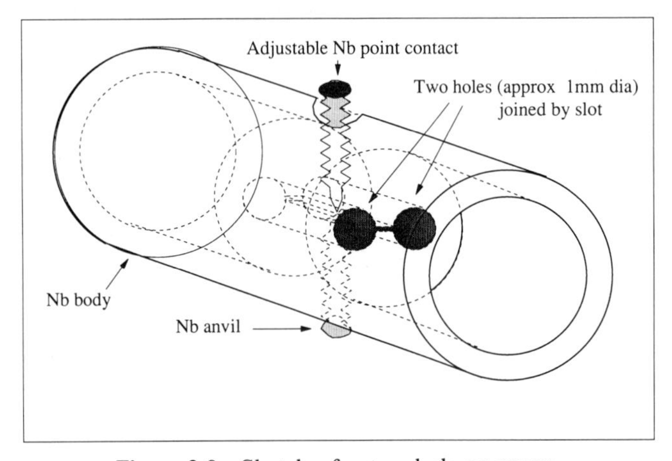

Figure 3.9 Sketch of a two-hole RF SQUID

3.4.2 The toroidal SQUID

More recently a manufacturer has developed the self-shielding concept even further by producing an RF SQUID of toroidal construction, based on an original design by Kamper and Zimmerman (1971). Two bulk pieces of niobium making a sliding fit allow the toroidal signal coil and tank circuit coil to be completely enclosed. A further advantage is that this geometry allows accurate calculation of the self- and mutual inductances of the various components. In the present version the weak link consists of a planar niobium tunnel junction with a semiconductor barrier, which is clamped between the two bulk pieces. The integrity of the toroid is only broken by the emergence through small holes of connections to the two coils from opposite ends. The RF coil is attached to a capacitor across the end of a low-capacitance coaxial line while the ends of the signal coil are welded to niobium screw terminals, to which superconducting contacts may be made. The sophisticated RF amplifier and audio-frequency modulation flux locking system supplied for use with such a SQUID allows a flux sensitivity of $5 \times 10^{-5} \Phi_0 \, \mathrm{Hz}^{-1/2}$ to be readily achieved, equivalent to an energy resolution of $10^{-29} \, \mathrm{J \, Hz}^{-1}$. Flux locking can be maintained in the presence of signals with a slew rate up to $10^7 \Phi_0 \, \mathrm{s}^{-1}$ and low-amplitude signals up to a frequency of 10^5 Hz may be measured. This device which, at the time of writing, represents the 'workhorse' SQUID, in use in the majority of all applications is far from being a state-of-the-art device, as we shall see in the following section and chapter, where we consider the performance which has been achieved with experimental SQUIDs. However it should be stressed that in most applications it is not intrinsic SQUID noise which limits system performance but extraneous interference. Some practical examples of this, together with suggestions for minimising unwanted signals, will be given in Chapter 7.

3.4.3 More advanced RF SQUIDs

An RF SQUID exhibits a power gain which is proportional to the pump frequency ω_0 (see equation (3.15)). The vast majority of devices have employed pump frequencies in the range 10–30 MHz, although commercial systems operating at 200 MHz may now be purchased. Only a number of experimental systems have been run at much higher frequencies, up to at least 90 GHz (Silver and Sandell 1983). Hollenhorst and Giffard (1979) have summarised the results on systems of this type and have shown that the basic frequency dependence is verified. It is interesting that the results obtained by the group at Sussex University are the only ones which significantly improve on the theoretical classical limit to single-junction SQUID performance (Prance *et al* 1981). We shall take up this point in greater detail in Chapter 8.

The power gain is only one aspect in assessing the overall sensitivity of a SQUID system. Thus, as we have seen, the first stage amplifier immediately following the device itself is probably the dominant noise source in most situations. At present noise temperatures of RF amplifiers operating up to about 500 MHz at room temperature are typified by an approximately frequency-independent value of around 100 K. Semiconductor amplifier noise temperatures rise rapidly at higher frequencies. There are various rather expensive low-noise alternatives such as cooled parametric amplifiers, and even maser amplifiers have been used (Pierce *et al* 1974). Alternatively, considerable success has been reported using liquid-helium-cooled GaAs FET amplifiers. For example, noise temperatures as low as 1 K have been reported at 440 MHz (Prance *et al* 1981). Figure 3.10 shows a block diagram of the detection system electronics described in this reference, which is capable of detecting large signals of many Φ_0 at frequencies up to 10 MHz in the linearised mode of operation.

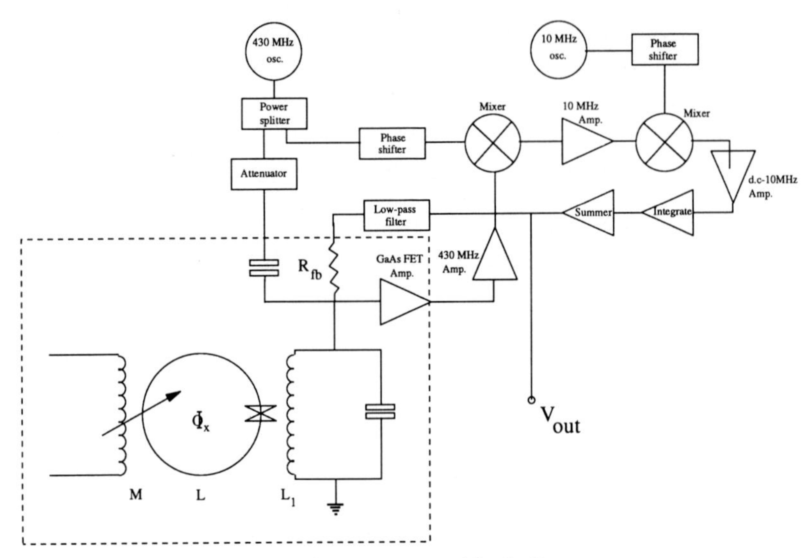

Figure 3.10 UHF SQUID block diagram

4

DC SQUIDs

4.1 Principles of Operation of DC SQUIDs

A DC SQUID consists of a ring of superconductor, interrupted by two Josephson junctions, preferably with at least similar properties (see figure 4.1). The introduction of a second weak link has the important effect of allowing a finite, time-averaged, direct voltage difference to be established across the junctions by a direct bias current. This arises because there is no strongly coupled superconducting short circuit across them, unlike the case of the single junction in an RF SQUID. (Although in principle the single junction could be DC voltage biased by a linear flux ramp applied to the ring, this is not a technique which has been much used in practice). Traditionally the operation of DC SQUIDs has been described in a phenomenological way and we shall first give such a description. For many years RF SQUIDs have dominated the applications field and although this has been mainly due to the greater difficulty of producing two similar junctions in a single loop it is also, one suspects, the result of a less than complete understanding of the DC device. It is only relatively recently that a numerical model of the two-junction system has been solved and even now a simple physical explanation of its operation is difficult to find, a situation which section 4.2 of this chapter attempts to remedy.

4.1.1 Phenomenological description of a DC SQUID

Let us assume critical currents $i_{1\,0}$ and $i_{2\,0}$ for the two junctions, when treated separately. The current–voltage characteristic of the DC SQUID, shown schematically in figure 4.2, looks at first sight just like that of a single junction with effective critical current $i_{1\,0} + i_{2\,0}$. However the total critical current is now found to be a periodic function of the magnetic flux applied to the superconducting ring containing the two junctions. The figure shows two $I-V$ curves corresponding to integer ($n\Phi_0$) and odd half-integer $(n+\frac{1}{2})\Phi_0$ values of the applied flux in units of Φ_0. The point to notice is that if the SQUID is biased by a constant current into the finite voltage regime, the time-averaged voltage appearing across the junctions is also a periodic function of flux. Thus the detection system of a DC SQUID is extremely simple, consisting at its most basic of just a low-noise audio amplifier and an adjustable constant-current source. This is to be compared with the much more complex RF circuitry required for the single-junction device. In real situations the DC SQUID detection system is a little more complicated than the simplest possible form since, just as for the RF device, low-frequency flux modulation and negative feedback can be used to provide linear response and a greater dynamic range (see figure 4.3).

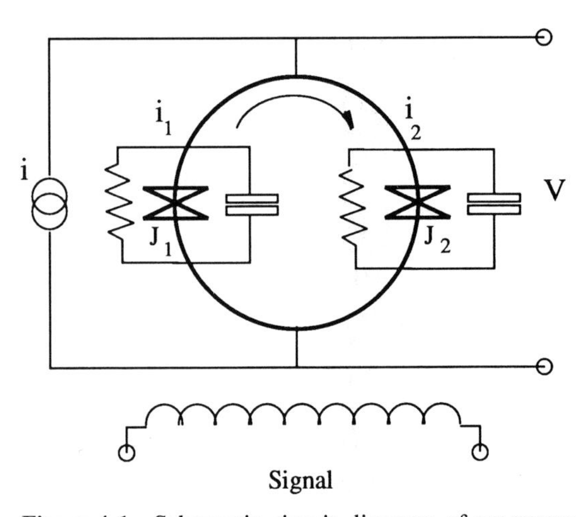

Figure 4.1 Schematic circuit diagram of DC SQUID

For about ten years, between the initial demonstration of SQUID behaviour by Silver *et al* (1966) and John Clarke's pioneering work in the mid-1970s (Clarke *et al* 1976), DC SQUIDs were almost totally ignored, in the mistaken belief that they were intrinsically low-frequency devices which were less sensitive than the RF SQUID competitors. This probably resulted from a lack of understanding of what is going on in the rather misleadingly named 'DC' SQUID. In fact, biasing the device into the finite voltage regime causes very high-frequency circulating alternating supercurrents to flow around the ring, as a result of the AC Josephson effect in the junctions. Since a typical bias voltage in the optimum situation is around 100 μV the supercurrents flow at frequencies of around 50 GHz, higher than that of almost any RF SQUID.

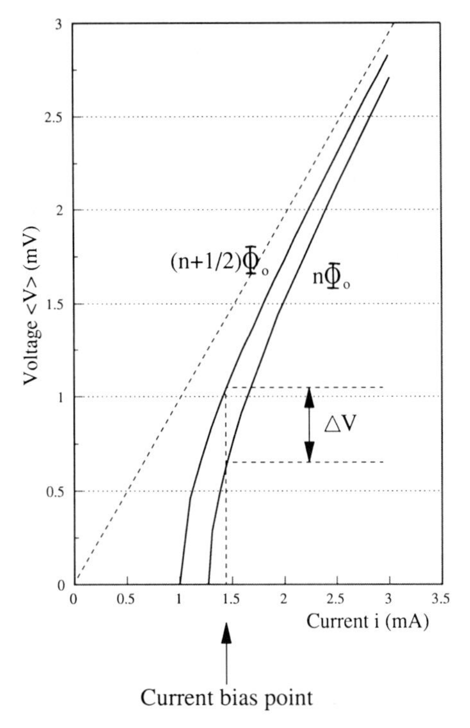

Figure 4.2 Variation of DC SQUID *I–V* characteristic for two applied flux values

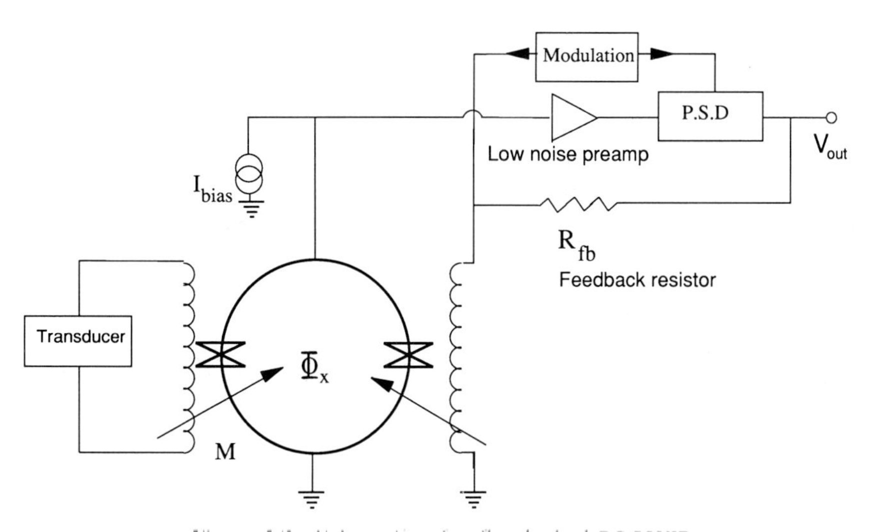

Figure 4.3 Schematic of a flux-locked DC SQUID

The basic mode of operation of the DC SQUID can be explained at least qualitatively by treating the device as possessing three distinct, but inseparably connected components: an oscillator, a parametric amplifier and a microwave detector. The oscillator consists of the two junctions biased at finite voltage so that an oscillating supercurrent at the Josephson frequency is driven round the loop. The second, parametric amplifier element results from the fact that the amplitude, phase and frequency of the circulating current depend periodically on the flux applied to the ring. Finally the amplitude of the circulating supercurrent is detected by the response of the junctions to it as described in section 2.4, so that the $I-V$ characteristic of the paralleled junctions is affected in a measurable way as the applied flux Φ_x is changed.

With circulating currents flowing at such high frequencies it is clear that the lumped circuit impedances of the ring and the junctions will become increasingly important if they are not to limit the performance of the SQUID. We shall see later that the loop inductance and the junction capacitances should all be minimised for an optimum design.

4.1.2 The DC SQUID as a parametric amplifier

The extreme non-linearity of the Josephson effects means that analytical results are few and numerical calculations are required to model accurately the behaviour of SQUIDs. Fortunately it is again possible here, as it also proved for the RF SQUID, to make use of a linear approximation in order to estimate the power gain of a DC device. A vital parameter for which we require an estimate is the maximum value of circulating supercurrent i_s which the ring can support. For small values of i_1 and i_2, i_s will be set by the smaller of these two. A simple argument suggests that another limit exists for large critical currents. Consider a DC SQUID biased by a direct current $i > i_1 + i_2$, so that a time-averaged voltage $\langle V \rangle$ exists across the junctions. The phase difference across each junction increases with time according to the second Josephson equation (2.11). If Φ_x is equal to to $n\Phi_0$ the phase differences of the two junctions (θ_1 and θ_2) evolve in step with one another. No circulating AC supercurrent flows round the ring. Changing the applied flux introduces a relative offset between the junctions and as a result a time-dependent circulating current flows, whose frequency is also determined by $\langle V \rangle$. If each junction's resistance is R and the loop inductance is L, above a frequency $f \sim R/\pi L$ the amplitude of the circulating supercurrent will be attenuated by the increasingly significant impedance of the loop inductance. The change in voltage resulting from a flux change $\Phi_0/2$ is

$$i_s R \sim \Delta V < \Phi_0 R/\pi L \qquad (4.1)$$

where the relation between ΔV and $\Phi_0 R/L$ derives from the AC Josephson effect, using the upper limit for the frequency. Thus the maximum circulating current, $i_s \sim 2\Phi_0/\pi L$, is that value which will apply approximately $1\Phi_0$ to the ring. For most realistic situations this limit will apply since $\Phi_0/L < (i_1 + i_2)$. Remember, though, that this argument is over-simplistic since it assumes a linear behaviour for the Josephson junctions whereas these devices are characterised by extreme non-linearity. Nevertheless it serves to give a physical insight to the limit on circulating current amplitude which is predicted by more detailed numerical calculations and confirmed by experiment. The estimate of i_s allows us to calculate the power gain of the DC SQUID if it is regarded as a parametric amplifier of the upconverter type. For a signal $\frac{1}{2}\Phi_0 \sin \omega t$, chosen since it is the maximum amplitude which will produce an output at the same, rather than a harmonically multiplied, frequency, then provided that $\omega \ll R/L$ the available input power is $\omega\Phi_0{}^2/8L$ whereas the available output power is less than

$$i_s \delta V/2 \sim i_s{}^2 R/2 \sim \Phi_0{}^2 R/\pi^2 L^2$$

(where $\delta V \sim i_s R$ is the change in $\langle V \rangle$ produced by a change of circulating current from 0 to i_s) so that the power gain P can be written

$$P < 8R/\pi^2 \omega L. \tag{4.2}$$

As suggested above, $\omega_p = R/L$ is the maximum ring current angular frequency and may be regarded as the pump frequency of the SQUID. Thus $P < 8\omega_p/\pi^2\omega$ and an upper limit to the power gain is set by the ratio of the pump frequency of i_s to the signal frequency. This is just the limitation on the power gain of a parametric upconverter set by the Manley–Rowe relationships.

4.1.3 Optimisation of DC SQUID performance

The arguments of the previous section allow us to predict the parameters required to produce an operating device which has as large a gain and available signal bandwidth as possible. (The third general requirement, low noise, will be dealt with in some detail in section 4.3.) To maximise the gain the resistance R should be as large as possible. For Josephson tunnel junctions the product $i_1 R$ is limited by the superconducting energy gap 2Δ (see section 2.5.1).

Since the optimum value of the critical currents should not fall below Φ_0/L if the available circulating current is to be kept at its maximum, this condition demands that

$$R \sim 2\Delta L/\Phi_0 e. \tag{4.3}$$

As an example, for a loop inductance $L \sim 1$ nH and niobium electrodes, for which $2\Delta/e \sim 2.8$ mV, we would require $R > 10\ \Omega$. The discussion of the tilted washboard analogue of DC SQUIDs, presented in Appendix B.1, introduces another limitation. If the junctions are insufficiently damped the transition from the zero to finite voltage regime is not reversible but shows hysteresis. This is a result of the junction capacitance C being too great (see section 2.2 and figure 2.3(*b*)). Once the junctions are in the finite voltage regime (where the phase is evolving with time) the capacitance, which is analogous to inertia in the washboard model, maintains a time-evolving junction phase difference even when the current bias is reduced below the level which caused them to switch first into the finite voltage state. This hysteretic behaviour has the unfortunate consequence of introducing high noise levels into the SQUID, via the sharp switching transitions which small transients or interference can bring about. Irreversible behaviour is avoided if the damping is adequate, requiring that the inequality

$$\beta_c = 2\pi i_1 R^2 C/\Phi_0 < 1 \tag{4.4}$$

be satisfied. Combining (4.3) and (4.4) leads to the condition

$$C < i_1 \Phi_0 / 8\pi\Delta^2. \tag{4.5}$$

For a tunnel junction there is also a relationship between the critical current density j and the capacitance c per unit area, for a given insulating barrier. The tunnelling supercurrent density depends exponentially on barrier thickness t whereas c varies as t^{-1}. The above condition (equation (4.5)) on the capacitance can be rewritten in terms of the ratio j/c. Perhaps surprisingly, the hysteresis problem can be minimised by making the barrier as thin as possible, even though this also maximises the capacitance. It is just that the supercurrent density increases faster with decreasing thickness than does c. Thus the best junctions have the highest possible current density, provided that the barrier can be made continuous and uniform. The techniques for achieving high current density have been developed as a result of the extensive Josephson logic development programmes undertaken by IBM and other large industrially oriented labs. These will be considered in some detail in Chapter 6.

4.2 Quantitative Treatment of DC SQUIDs

Much of the behaviour of DC SQUIDs can be understood from the qualitative picture given in the previous section, but it is also quite straightforward to calculate quantitatively how the devices will behave, at least in the classical limit. Let us simplify the argument by assuming that the device is ‘symmetric’, meaning that both arms of the superconducting ring are identical with respect to the critical current and normal-state resistance of each junction, and the distribution of self-inductance. A somewhat asymmetric SQUID exhibits only slightly different properties but considerably more complex calculation and presentation of the results is required. Figure 4.1 shows a schematic circuit diagram of a current-biased DC SQUID. When the flux applied to the ring is not an integral number of quanta, the bias current i does not divide equally between the two arms. If the currents flowing through each arm are i_1 and i_2 then clearly $i = i_1 + i_2$. This asymmetry can be expressed in terms of a current j which circulates around the loop so that

$$j = (i_1 - i_2)/2. \tag{4.6}$$

Each junction is characterised by the circuit parameters shown in figure 4.1, consisting of an ideal Josephson element with critical current i_c and sinusoidal current–phase relationship, shunted by a normal resistance R and a capacitance C. We will assume that $C = 0$ initially. Thus the current flowing through each arm consists of a supercurrent and a normal component in parallel:

$$i_1 = i_c \sin \theta_1 + V_1/R \tag{4.7}$$

$$i_2 = i_c \sin \theta_2 + V_2/R. \tag{4.8}$$

The AC Josephson effect provides us with a relationship between the time-dependence of the phase difference θ and the voltage V, whereas the DC Josephson effect relates the supercurrent component to the same phase difference. Also the total voltage drop across the SQUID V is the sum of the voltage drop across the junction plus that across the inductance of each arm, given, of course, by the term $L\, \mathrm{d}i/\mathrm{d}t$. Combining these expressions yields a set of four coupled normalised equations:

$$j = \Phi_0(\theta_1 - \theta_2 - 2\pi\Phi_x)/2\pi L \tag{4.9}$$

$$V = (\mathrm{d}\theta_1/\mathrm{d}t + \mathrm{d}\theta_2/\mathrm{d}t)/2 \tag{4.10}$$

$$\mathrm{d}\theta_1/\mathrm{d}t = i/2 - j - \sin \theta_1 \tag{4.11}$$

$$\mathrm{d}\theta_2/\mathrm{d}t = i/2 + j - \sin \theta_2. \tag{4.12}$$

The non-linearity, arising from the sinusoidal terms, means that these four equations have no general analytic solutions. However, they may be readily integrated numerically to provide time-averaged values for voltage and circulating current as functions of applied bias current and flux. Some typical current–voltage characteristics computed from this model are given in figure 4.2, while figure 4.4 illustrates how the circulating current j is a periodic function of applied flux (with period Φ_0). It is interesting at this point to notice that the maximum amplitude of j is only Φ_0/L, regardless of the value of $\beta = 2\pi L i_c/\Phi_0$. This numerical result may be given a physical basis by the following argument. When the SQUID is biased at a non-zero voltage, the phase difference across each junction increases with time. Now the relative phase differences between the two junctions is set by the applied flux, as is apparent from equation (4.9). It is energetically unfavourable for the relative phase difference to stray from the mean value set by the applied flux by more than $\pm\pi$. This range of relative phase 'wobble' allows a circulating current of amplitude only $\pm\Phi_0/2L$ to flow. In a subsequent section this argument will be amplified and also, in Appendix A.3, it will be discussed in terms of an analogue represented by two balls coupled together by a light spring and rolling on two inclined adjacent corrugated 'washboard' surfaces, whose relative phase can be adjusted.

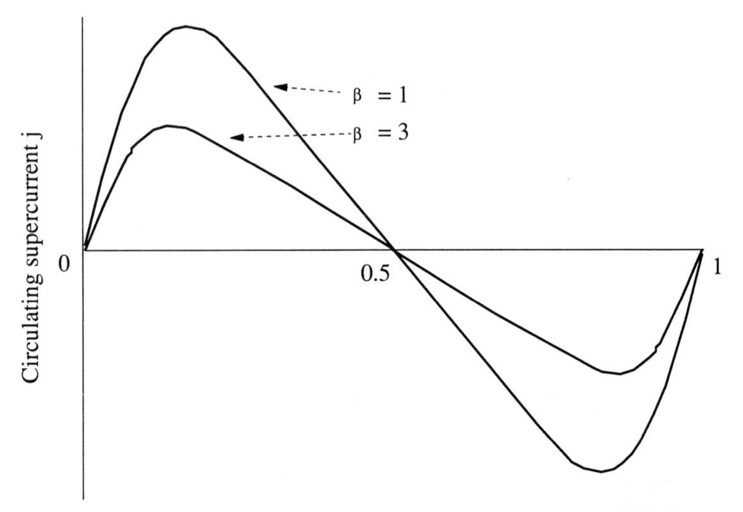

Figure 4.4 Plot of mean circulating supercurrent versus applied magnetic flux (Φ_x) in units of Φ_0

Figures 4.5 and 4.6 show some typical computed curves for the highly symmetric case we are considering. It is clear that the voltage and circulating current oscillate in time about their mean values with a frequency which is proportional to the mean voltage itself. This, of course, is just the Josephson frequency $\omega = 2eV/\hbar$. Thus the mean voltage drives the circulating supercurrent, and this in turn affects the relative phase difference across the two junctions, in a way which depends on the applied flux. This adjustment of the relative phase difference in turn reacts back on the time evolution of the voltage across the junctions so that its time-averaged value is changed by an amount dependent on the applied flux.

4.2.1 The gain of a DC SQUID

In section 3.3 the gain of an RF device was shown to be proportional to the

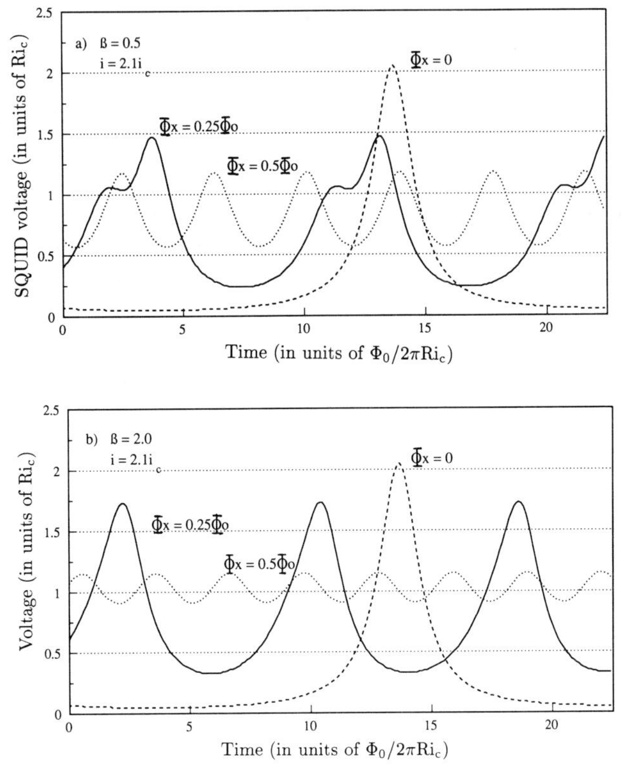

Figure 4.5 DC SQUID voltage versus time for three different values of the applied flux with (*a*) $\beta = 0.5$, and (*b*) $\beta = 2.0$

pump frequency ω_{RF}. The rather simplistic treatment of section 4.1.2 hints that in an analogous way the gain of the DC SQUID is proportional to the frequency of the alternating circulating current and voltage. From figure 4.6 it is clear that when $\Phi_x = n\Phi_0$ the circulating current is zero at all times. The phase differences across each junction just increase exactly in step with no wobble. Since the junctions are in the finite voltage regime a sharp voltage spike appears, its width being determined by the circuit parameters L/R. The optimum bias point for the junctions occurs when the mark-to-space ratio for these voltage spikes is about one, since in this case the time-averaged change in voltage produced by a given change in applied flux will be a maximum. This occurs for the condition

$$\omega = 2\pi V/\Phi_0 = 2\pi R/L$$

so that ω is effectively the maximum frequency at which a particular device can be operated, yielding an estimate of the power gain g. The available output power δP corresponding to an input flux change $\delta\Phi_x$ in a bandwidth df is

$$\delta P = \omega(\delta\Phi_x)^2/L$$

so that

$$g = 2L\delta P/(\delta\Phi_x)^2\, \mathrm{d}f = \omega/\mathrm{d}f. \tag{4.13}$$

Comparison of this result with equation (3.15) strongly suggests that the frequency of the circulating supercurrent in a DC SQUID is analogous to the pump frequency of an RF SQUID, confirming the argument of section 4.1.2.

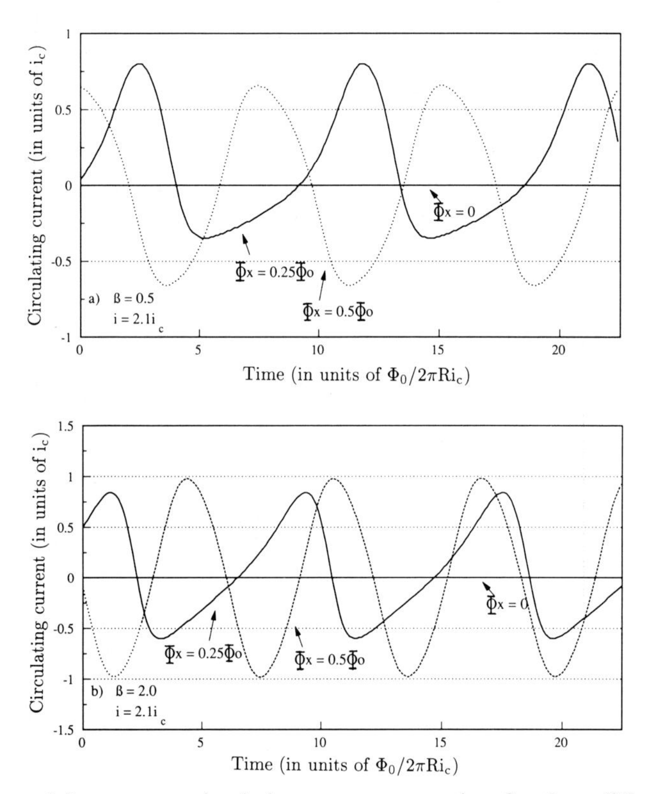

Figure 4.6 DC SQUID circulating current versus time for three different values of the applied flux with (*a*) $\beta = 0.5$, and (*b*) $\beta = 2.0$

4.3 Noise in DC SQUIDs

The claim was made in section 3.3.1 that in an RF SQUID external circuit noise is almost always much more significant than its intrinsic noise. For DC devices the situation is usually reversed. In this case the dominant sources of noise are, in a classical analysis, Nyquist voltage and current fluctuations associated with the normal state junction resistance. For a ring inductance L and two symmetrical junctions with critical currents i_c and resistances R the spectral density of the noise voltage appearing across the junctions is

$$S(V) = 4kTR/2.$$

In addition there will be current fluctuations in the ring with spectral density

$$S(i) = 8kT/R$$

and the simplest assumption is that these two noise sources are uncorrelated. This will not be the case in general, since circulating noise currents will influence the voltages appearing across the junctions and vice versa, but for simplicity we will ignore any extra contribution from these cross-correlations. The flux spectral noise density $S(\Phi)$ is simply related to both the minimum detectable energy in unit bandwidth δE and to $S(V)$ and $S(i)$:

$$\delta E = S(\Phi)/2L = [S(V)/(\partial V/\partial\Phi)^2 + S(i)/(\partial i/\partial\Phi)^2]/2L. \tag{4.14}$$

The essential non-linearity of a SQUID's response means that to evaluate the partial derivatives in the above expression requires numerical solution of the differential equations (4.9)–(4.12), but now including random noise sources of current and voltage fluctuations. Such an analysis has been given by Tesche and Clarke (1977) and more recently the problem has been solved for the situation of significant capacitance shunting the point contacts (Voss and Webb 1981). However for our purpose an approximate solution of the equations can be given which both yields results which are correct to within a small numerical factor and allows the dependence of the limiting energy sensitivity on various circuit parameters to be seen.

In section 4.1.2 it was argued that the maximum change in critical current which can be produced by an external flux was $\Phi_0/2L$ and this required a change of applied flux of $\Phi_0/2$. Thus $\partial i/\partial\Phi \sim 1/L$. Similarly

$$\partial V/\partial\Phi \sim (\partial V/\partial i)\,(\partial i/\partial\Phi) = R/2L.$$

Substituting these approximate values back into equation (4.14) yields the minimum detectable energy δE, set by thermal fluctuations in the SQUID ring of

$$\delta E \sim 8kTL/R. \tag{4.15}$$

This has the satisfyingly simple physical basis that a SQUID possesses two degrees of freedom (voltage and current) each of which is associated with $kT/2$ of thermal noise. This is distributed as white noise throughout a bandwidth $\sim R/2\pi L$.

4.3.1 Zero-point fluctuations

The above expression (equation (4.15)), derived from numerical calculations based on the resistively shunted junction model, suggests that if a SQUID can be operated at a sufficiently low temperature the limiting energy resolution may be lowered without limit. This is intuitively unreasonable since, as T is reduced, it would imply at some point the occurrence of some violation of the uncertainty principle relating energy resolution to measurement time. Several workers recognised this problem and a number of *ad hoc* modifications to the model were introduced to try to resolve the paradox. It was initially suggested that shot noise processes in the weak links would set a temperature-independent limit to the noise. Although appealing, this suggestion seemed to be in conflict with an experimentally observed absence of shot noise in tunnel junctions. Others pointed out that the Nyquist noise formula was only a high-temperature approximation to the appropriate expression for the noise current spectral density of an ensemble of quantised oscillators (Gallop and Petley 1976):

$$S(i) = 4\hbar\omega/R\{[\exp(\hbar\omega/kT) - 1]^{-1} + \tfrac{1}{2}\} \tag{4.16}$$

so that now $S(i)$ is dependent on the frequency ω when $\hbar\omega > kT$, but reduces to the classical white noise spectral density in the limit that $\hbar\omega \ll kT$. This modification allows the Heisenberg uncertainty relationship to determine the behaviour in the low-T limit. Some measurements have demonstrated the reality of the zero-point fluctuations which equation (4.16) predicts (Koch *et al* 1981). More recently it has become clear that the approach which this expression embodies is only a first attempt at coping with the real quantum nature of SQUID devices, and *ad hoc* tinkering with classical calculations cannot give a convincing answer to the problem. Together with the quantum tunnelling question this problem motivated developments towards a true quantum mechanical treatment of SQUIDs, a task in which considerable progress has been made over recent years, although it remains to be solved in all details (see Chapter 8).

4.3.2 Other noise sources

The direct dependence of δE on ring inductance L led to the production of DC SQUIDs with smaller and smaller loop areas. This development coincided with the arrival of thin film photolithographic techniques which allowed very small-scale devices to be fabricated in large numbers. It became apparent, particularly from the work of the IBM group, who used loop inductances as low as a few pH, that the expression (4.15) was confirmed by experiment. However it was also found that the white noise which it predicts only existed above some cut-off frequency ω_c. Below this the spectral density increased roughly as $1/\omega$. In general, as the white noise was reduced ω_c moved to higher and higher frequencies so that for some of the best SQUIDs $\omega_c \sim 100$ kHz, setting severe limitations on their practical usefulness. A number of suggestions have been put forward to explain this low-frequency noise, notably that temperature fluctuations generate critical current changes in the junctions, which appear as noise. However the expected critical current temperature dependence is not seen in the $1/\omega$ noise contribution. Work by Rogers and Buhrman (1984) indicates that tunnel barrier height fluctuations across the area of the junction may simulate a $1/\omega$ noise spectrum. These fluctuations may arise from electron trap inhomogeneities so $1/\omega$ noise may be reduced as the barrier quality improves. It appears to be less prominent in point-contact devices than in thin film tunnel junctions.

Recently SQUIDs with relatively large values of L (0.1 nH) and C (0.1 pF) have been cooled below 1 K using dilution refrigerators to show that equation (4.15) is obeyed down to at least 50 mK, provided that the quantum limit has not already been reached at higher temperatures and that 'hot-electron' effects are avoided. The latter occur if the shunt resistors are physically too small to allow thermalisation of the conduction electrons in transit through the resistor. The effect manifests itself as excess Johnson noise corresponding to the higher temperature of the non-thermal equilibrium current carriers and the solution has been shown to make the thin film resistors as large as possible (Wellstood *et al* 1989).

4.4 Practical DC SQUIDs

The first SQUID ever described was of the DC type (Silver *et al* 1966), using two large-area Josephson tunnel junctions incorporated into a thin film superconducting loop of very low inductance. Following this initial demonstration, very little was heard of double-junction devices for more than a decade, except for a brief appearance of the Clarke 'SLUG' (Clarke 1966). This consisted of an oxidised niobium wire on whose surface two scratches were made a few millimetres apart. Then a blob of molten solder was formed around the region containing the scratches and allowed to solidify. By some poorly understood process, Josephson junctions of the point-contact type are sometimes formed between the superconducting solder drop and wire. External flux is coupled into the loop, formed by the small cross-sectional area occupied by the oxide between solder blob and wire, by passing a current along the Nb wire. This rather pragmatic device was used by a number of workers and good low-noise performance was reported. However such poorly characterised devices were clearly unsuited for systematic study aimed towards a better understanding of operation, and hence improved performance, so they have been replaced by other types. Several groups have investigated DC SQUIDs formed from two pieces of bulk niobium, electrically insulated but held rigidly together. The two pieces, which are shaped to take a solenoidal signal coil, are bridged by two adjustable point-contact junctions. Very high sensitivities have been reported (see for example Stevens (1984)), but again such devices are not usually found to be stable, particularly if cycled between ambient and liquid helium temperatures, so consequently they have been little used except to determine the potential performance of DC SQUIDs. Figures 4.7(*a*) and 4.7(*b*) show the schematic form of the SLUG and point-contact DC SQUID.

The major breakthrough, which brought DC SQUIDs back into serious consideration as practical devices, was made by Clarke and his group at Berkeley, over a number of years from about 1975. They produced a DC SQUID using two Nb–PbIn tunnel junctions, of relatively small area (10^{-2} mm^2). These were incorporated in a thin film superconducting lead ring evaporated onto the surface of a quartz tube (figure 4.7(*c*)). To prevent hysteresis in junctions with these intrinsic properties it was necessary to shunt them with low resistances (0.42 Ω gold strips). External signal coupling to the ring was made through a signal coil inserted in the quartz tube. The detection system used for this SQUID was also novel, and it is described in more detail in the following section. The importance of this work was that the performance achieved was considerably better than that available from commercial RF SQUIDs, particularly in the low-frequency noise performance in the low-frequency region, where $1/f$ noise dominates both types of SQUID. Furthermore this SQUID proved reliable and robust and has been used in a number of real applications, particularly in the field of geophysics (see section 5.7). Reported flux sensitivity was $10^{-5}\Phi_0\,\mathrm{Hz}^{-1/2}$ and this agrees well with theoretical estimates based on the assumption that the limiting noise in DC SQUIDs originates as Nyquist voltages and currents in the resistive shunts of the junctions. This SQUID not only proved to be a useful device, but also provided a suitable test platform for theoretical models of the DC SQUID, which enabled much greater understanding of its basic properties.

At the beginning of the 1980s the thin film Josephson junction technology developed by IBM for digital SQUID devices had reached the point where tunnel junctions of area as small as 1 μm^2 could be fabricated. The area, and hence the capacitance, of these junctions is of order 10^4 times smaller than that of the Clarke type and consequently the hysteresis parameter $\beta_c \ll 1$ so it is not necessary to provide additional shunt resistors across the junctions. This in turn removes a major source of intrinsic noise. A number of workers produced DC devices based on small-area tunnel junctions and thin film planar rings. The energy sensitivity per unit bandwidth demonstrated with these ranged up to $1.7\hbar$ (Planck's constant), in

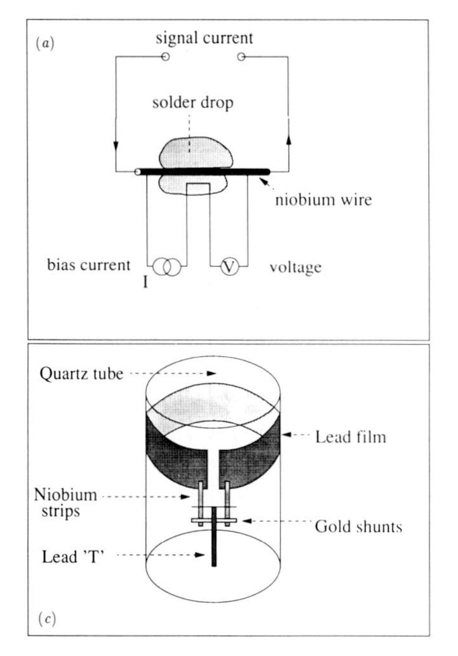

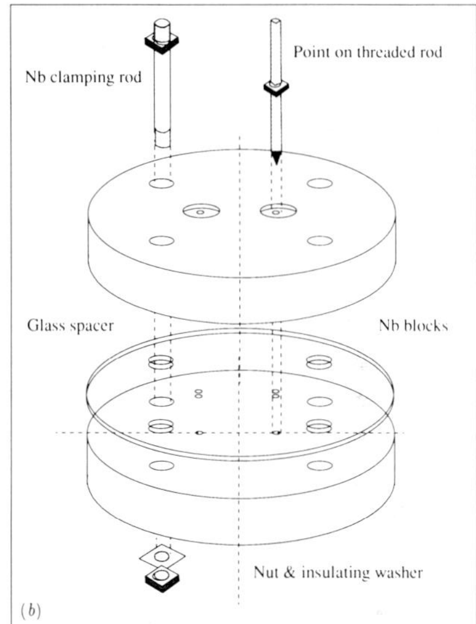

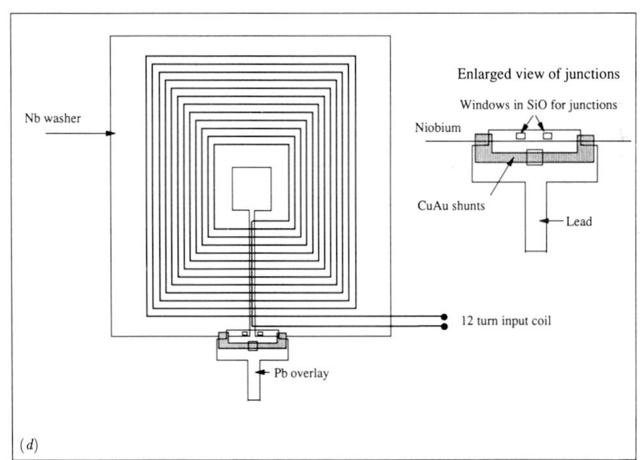

Figure 4.7 (*a*) A solder drop SQUID (also known as a SLUG); (*b*) construction of an adjustable point-contact DC SQUID. The two niobium blocks are insulated by a glass spacer. Only one of two points and one of four clamping rods is shown. (*c*) A simple tunnel junction DC SQUID; and (*d*) a thin DC SQUID with 'washer-coupling' input coil

agreement with the classical limit calculated from the generalised form of the Nyquist theorem (Awschalom *et al* 1989). The use of these SQUIDs in real applications is still anticipated with great interest since it may be expected that in principle quantum-noise-limited SQUIDs are capable of detecting single photons down to frequencies in the audio range. Although the number of applications for audio photon detectors is at present rather small it seems certain that much interesting physics could be done with such a device (see section 8.2.1). The first serious application of these SQUIDs has been reported (Ketchen *et al* 1989) in which the device is integrated using LSI fabrication techniques into a picosecond magnetic susceptometer which is able to detect the magnetisation of as few as 10^3 electronic spins in a sample of a dilute magnetic semiconductor. Laser pulses can be used to manipulate the magnetisation of such materials, allowing the SQUID to detect spin changes arising from picosecond pulses.

A restriction on the applications of these 'second-generation' DC SQUIDs is that they possess extremely small ring inductances, perhaps as low as 1 pH. This makes it difficult to couple an external flux signal efficiently into the SQUID ring. (It is of course a direct consequence of the small inductance that the energy resolution is so good (equation 4.15)). Furthermore these are, unlike the cylindrical Clarke SQUID, planar devices, with thin film rings, which again makes for coupling difficulties. Two solutions to these problems have been followed. Jaycox and Ketchen (1981) formed the SQUID ring from a thin film 'washer' with a small central hole of small inductance and a slot, across which the junctions were placed, a strip line configuration being laid over the slot to reduce the parasitic inductance. The outer diameter of the washer can be made large enough to provide tight coupling to a multi-turn (~ 100) coil without significant increase in L. In this way devices have been produced with input coil inductances of 800 nH, which nevertheless are tightly coupled ($k^2 = 0.86$) to a SQUID ring inductance of only 90 pH. Figure 4.7(c) shows a schematic view of a washer-coupled SQUID.

The second approach to coupling flux into planar SQUIDs takes up an idea originating with point-contact bulk-structure RF SQUIDs. Cromar and Carelli (1981) have produced a thin film SQUID loop with an inductance of 6 pH. This is well coupled to 68 loops in parallel, arranged around the perimeter of a square. These loops are further surrounded by a 20 turn input coil, which could be the secondary of a flux transformer, for example. This technique is reminiscent of Zimmerman's (1972) multi-hole SQUIDs (see section 5.3.2) where the total inductance of a large-area coil combination is reduced by connecting many loops in parallel. A coupling constant $k^2 = 0.43$ has been achieved by Cromar and Carelli (1981), giving a flux noise spectral density per unit bandwidth $S_\Phi^{1/2} = 1.6 \times 10^{-7} \Phi_0 \,\mathrm{Hz}^{-1/2}$. Clearly both this method and the 'split-washer' technique outlined above provide practical ways of adequately matching SQUIDs to external circuits.

4.5 DC SQUID Detection Systems

Although the need to produce two reasonably well matched Josephson junctions in the same ring means that DC SQUIDs are rather more difficult to manufacture than the RF variety, this disadvantage is offset to some degree by the much simpler detection electronics which the DC variety requires. Figure 4.3 shows the essential elements of the detection mechanism, which are as follows: first, a means of direct current biasing the junctions into the finite voltage regime, second, a low-noise amplifier to detect the voltage change across the junctions, produced by a change in the flux applied to the superconducting ring and, finally, some means of transforming the physical parameter to be measured into a flux change in the ring. Typically the short-term bias current stability must be better than 1 in 10^4. Battery-powered supplies are very suitable and have the added advantage of avoiding earth loop difficulties. Room temperature preamplifiers, to follow the SQUID, are available with noise figures as low as < 0.3 dB, optimised for a source impedance of $10^4\ \Omega$.

If the SQUID is biased with only a direct current while a linear flux ramp is applied to the ring, the voltage across the junctions varies periodically with flux, although the variation is not truly sinusoidal, becoming more cusp-like as the current bias is reduced and the mean voltage approaches zero (see figure 4.5). The current bias point which gives the system its maximum gain, that is the maximum value of the flux to voltage transfer ratio $\partial V/\partial\Phi$, varies as the applied flux changes. Optimally the bias point would be adjusted to give a small voltage across the junctions for each value of applied flux. This is obviously not a very practical arrangement and in most real SQUIDs a gain-linearising circuit, based on negative feedback, is included, very similar in principle to the flux-locking circuit used with RF SQUIDs (see section 3.2). In addition to the signal winding, this requires the provision of a second coil coupled to the superconducting ring. Flux modulation, of amplitude $\Phi_0/2$ at an audio frequency ω_0, is applied to this coil. The voltage output across the SQUID is amplified, phase-sensitively detected at ω_0 and then, after suitable filtering to prevent oscillation, fed back into this coil. Thus the SQUID bias point may be maintained by negative feedback, which ensures that the sum of the applied flux and fed-back flux is constant. Changes in feedback current are now proportional to changes in the applied flux, so the gain of the device has been linearised, making the SQUID more generally useful.

Since SQUIDs are such low-noise devices, considerable care must be paid to the first stage amplifier into which the junction voltage is fed. In most applications the bandwidth of the overall system is limited to the audio frequency range. A major problem lies in matching the very low output impedance of the two Josephson junctions (typically 1 Ω or less) to the normally very high input impedance (typically 100 MΩ) of the best room temperature FET amplifiers. Two rather different approaches have been followed. One method (Duret and Karp 1983) which has been demonstrated is to impedance match using a simple transformer. The transformer must be very carefully designed to achieve reasonable matching over such a large impedance range while still retaining a tolerable signal bandwidth. The second, and most developed, solution to this problem is, like so much in the field of DC SQUIDs, due to the work of the group led by John Clarke at Berkeley. They have used a novel approach by employing a series tuned circuit across the voltage terminals of the SQUID. The amplifier is fed from the tuned circuit capacitance and the impedance at resonance is that of the normal-state resistance of the paralleled junctions, multiplied by the Q value of the low-temperature circuit, which can be several thousand—even at frequencies as low as 100 kHz (figure 4.8). The resonant circuit serves another purpose in that the flux modulation associated with the linearising feedback is applied at the resonant frequency ω_0 of the circuit. The narrow-band nature of its response serves to isolate the SQUID from noise and interference at other frequencies. It does of course limit the signal bandwidth of the linearised device to around $df = \omega_0/2\pi Q$, which may only be of order 1 kHz. There is no obvious upper limit on ω_0. Up till now audio frequencies have been used, as phase-sensitive detectors are usually limited to this range. However, PSDs operating in the tens of MHz region are now available so that it seems that flux-locked DC SQUIDs with signal bandwidths up to around at least 1 MHz could be built, comparable in performance for noise, bandwidth and slew rate with the best RF devices.

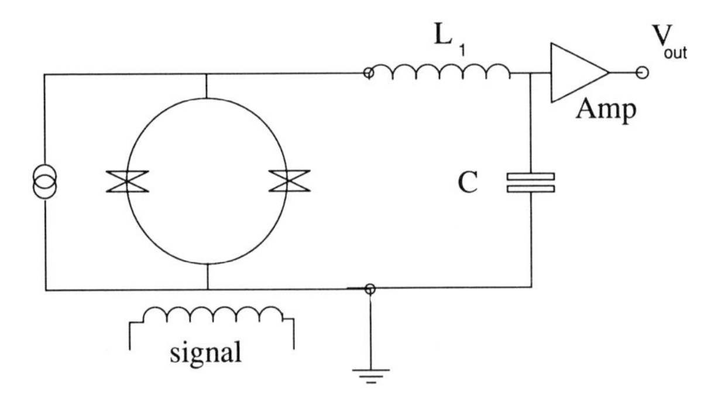

Figure 4.8 Schematic of a tuned-circuit DC SQUID readout

4.5.1 Self-oscillating DC SQUID detection system

With the requirement that large numbers of SQUID sensors be integrated into a single detection system, to be used for example in biomagnetic imaging systems, it has become important to simplify SQUID readout electronics as much as possible. A novel technique has been demonstrated which employs only two integrated circuits and no low-noise preamplifier, the entire unit including power supply being only the size of a matchbox (Muck and Heiden 1989). The essentially different feature of this readout technique is that hysteretic Josephson junctions are employed and a series-connected resistance R_F and inductor L_F across them provides feedback which leads to a 'self-oscillating SQUID'. The frequency of this relaxation–oscillation state is given by

$$f = (R_F/L_F)\log_e[(i_b - i_{cmin})/(i_b - i_{cmax})]$$

where i_b is the bias current and i_{cmax} and i_{cmin} are the currents at which the SQUID switches out of, and back into, the zero voltage state (see figure 2.3(*b*)). Since these latter switching currents are functions of the magnetic flux Φ_x applied to the ring the frequency of oscillation will be also. The advantage of this arrangement is that the mean voltage across the SQUID is of the order of 10 mV, so the need for a low-noise preamplifier is eliminated and the change of frequency with applied flux $df/d\Phi_x$ may be measured simply with a standard frequency-to-voltage converter chip, the output of which may be fed back to the bias current supply to linearise and flux lock the SQUID. The use of hysteretic junctions might be expected to compromise the noise performance of the detection system but the flux noise for a SQUID made using Nb nanobridges is at the level of $10^{-5}\Phi_0$ $(\mathrm{Hz})^{-1/2}$ and white above 1 Hz, thus sufficiently low for most applications.

5

Applications of SQUIDs to Analogue Measurements

5.1 Introduction

The macroscopic theory of superconductivity indicates that SQUID behaviour depends in a fundamental way on magnetic flux quantisation. The output state of a SQUID is a periodic function of applied flux, but the absolute origin of the flux scale cannot be identified. This means that the device is essentially a sensor of flux changes. There are few applications which require only flux changes to be measured, but by coupling a suitable transducer to the input of a SQUID a very wide range of physical parameters may be sensed with the highest sensitivity.

Thus, most obviously, if the area of the input coil is known, a measured flux change may be converted to a flux density (magnetic field) change. This rather trivial example certainly does not exhaust the possibilities. We shall see in this chapter that SQUIDS have been used to measure electronic and nuclear magnetisation, field gradients from a wide variety of sources such as submarines, human muscular and nervous activity or the superfluidity of ^{3}He, temperature, displacement, DC and AC voltage and current, charged particles, magnetic monopoles, quarks and contamination from baked bean cans. We will treat these applications according to the physical parameter being measured, but a description of a number of SQUID uses will be deferred to Chapter 8.

5.1.1 *Figure of merit for SQUID applications*

Before describing some specific applications, there is one point which needs to be cleared up. In Chapters 3 and 4 the intrinsic sensitivity of both RF and DC SQUIDs was estimated. The possibility of producing quantum-limited devices has been realised, at least for the DC variant. However as equations (3.20) and (4.15) show, to produce the most sensitive SQUIDs requires that the loop inductance L should be minimised. (The quantum-limited DC SQUID of Ketchen *et al* (1989) had a value of $L \sim 1$ pH.) There are severe problems in coupling any external physical quantity into such a small inductance. At present, most real applications have not been able to employ SQUIDs of the lowest intrinsic noise due to these coupling difficulties. In order to compare the suitability of different devices for the same application we require some parameter which incorporates the effectiveness with which the coupling problem has been solved. The minimum detectable input energy change in unit bandwidth δE has proved to be a suitable figure of merit. For a SQUID with an equivalent total flux noise per unit bandwidth $\delta\Phi$ with a signal coil of self-inductance L_2, having a mutual inductance $M = k(LL_2)^{1/2}$ with the SQUID ring:

$$\delta E = \pi\delta\Phi^2/2k^2L. \tag{5.1}$$

The best value of δE known to the author at the time of writing is 1.7×10^{-34} J Hz^{-1} (Awschalom *et al* 1989), just a factor of 1.7 greater than the uncertainty principle limit. Ketchen *et al* (1989) have produced a SQUID with a much larger inductance loop ($L = 0.1$ nH) and 0.14 pF shunt capacitance which exhibits sensitivity close to the quantum limit when cooled to 0.3 K.

5.2 Magnetic Field Measurements

Any change of ambient field in which a SQUID is situated causes the flux linking the superconducting loop to change, so that SQUIDs can detect variation of the magnetic field without the need for any additional transducer attached to the input. However such a 'naked' SQUID has almost never been used, for good practical reasons. First, any bulk superconducting structure which it possesses will very adequately shield out, or at least seriously distort, any field change. Secondly, in practice SQUIDs must be operated within the relative magnetic stability of a superconducting shield, and any external field changes to be measured should then be coupled into the superconducting ring by means of a DC transformer. This device is shown schematically in figure 5.1 and consists of at least two coils connected in series, both usually superconducting (although not necessarily so, if field changes only above some arbitrarily low frequency are of interest). One of these coils, the secondary, is tightly coupled to the superconducting loop of the SQUID. If we assume that both coils are strongly superconducting then the total flux linking the coil combination is quantised, and therefore constant in time. However any change in the field applied to any of the external coils (which collectively form the primary) will change the total flux applied to the primary–secondary combination, which in turn will produce a change in the circulating persistent current. If the self-inductance of the secondary coil is L_2 and that of the primary L_1, with n turns of cross-sectional area A, a field change of δB will require that the supercurrent changes by an amount δi, where

$$\delta i(L_1 + L_2) + \delta B \cdot A = 0. \tag{5.2}$$

We have here neglected the inductance of the SQUID itself, as well as that of any associated shields. It might be expected that the sensitivity to magnetic field would be maximised by making the primary coil as large as possible. For a mutual inductance M between secondary and SQUID, the flux change at the SQUID is

$$\delta\Phi = M\delta i = nMA\delta B/(L_1 + L_2). \tag{5.3}$$

There is a relationship between A and the primary inductance, which depends on the coil's precise geometry. To be specific, we will assume that both coils approximate to the ideal of a long solenoid, for which the self-inductance is

$$L_1 = \mu\mu_0 n^2 A/l \tag{5.4}$$

where l is its length, μ being the relative permeability of the space within the coil. By differentiating $\delta\Phi$ with respect to A it is easy to show that maximum flux transfer from primary to secondary occurs for the condition $L_1 = L_2$. This at first sight surprising result is related to the maximum power transfer theorem in conventional electronics, for sources and loads with real impedance.

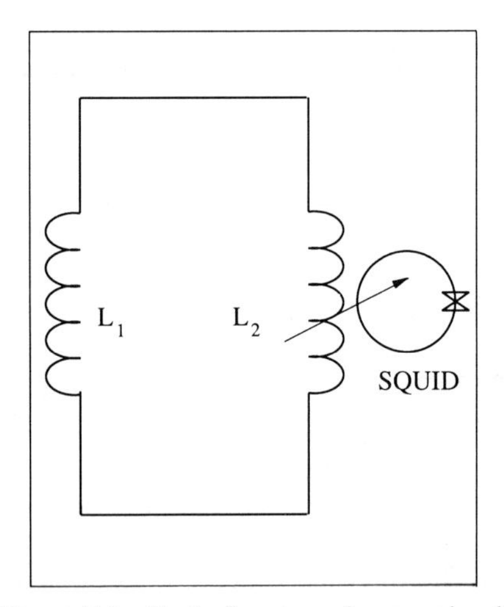

Figure 5.1 Basic flux transformer circuit

5.3 Field Gradient Measurements

In perhaps a majority of applications the primary of the flux transformer does not consist of a single simple coil as described in the previous section. Consider instead the arrangement shown in figure 5.2 in which there are two primary coils in series with *n* turns each, identical except that they are wound in opposite senses. Now a spatially uniform external field change will produce equal but opposite flux changes in the two coils, summing to

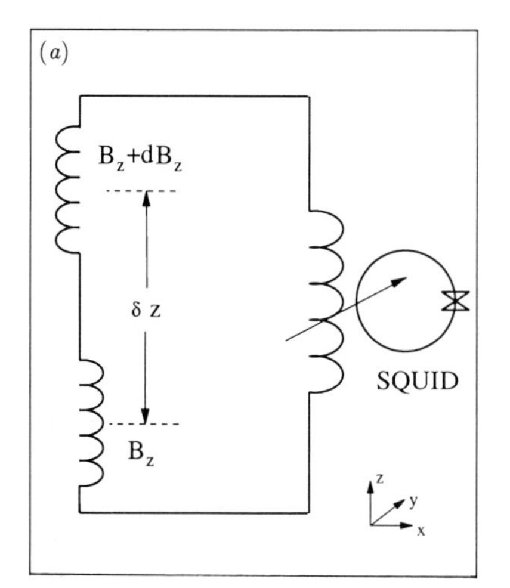

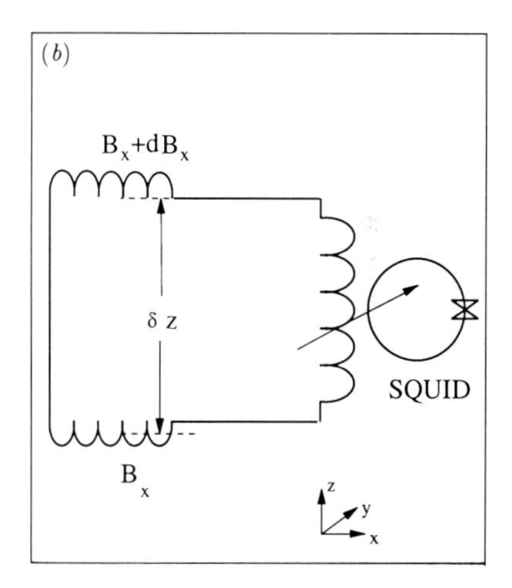

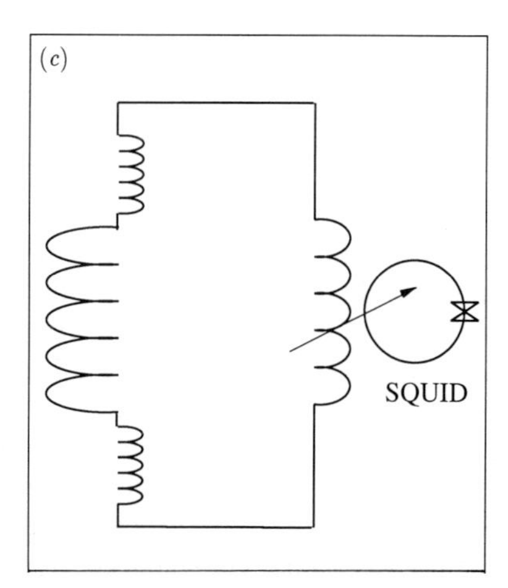

Figure 5.2 (*a*) A first-order diagonal gradiometer (primary coils wound in opposition); (*b*) a first-order off-diagonal gradiometer measuring the dB_x/dz component of the tensor (primary coils wound in opposition); and (*c*) a second-order diagonal gradiometer with a large centre coil; two small end coils constitute primary winding

zero with no change in the circulating current or the flux applied to the SQUID ring. If however the field change has a non-zero spatial gradient in the direction separating the coils, this cancellation will not be exact and an output signal will be detected. For the case shown in the figure in which the coils are separated by a distance δz in the z direction the arrangement is sensitive to changes in dB_z/dz. The flux change at the SQUID is

$$\delta\Phi = MnA(\mathrm{d}B/\mathrm{d}z)\delta z/(L_1 + L_2) \tag{5.5}$$

where A is the vector representing the coil area, directed perpendicular to its plane. The gradient of a vector quantity such as magnetic field is of course a 3×3 tensor quantity, with nine components in all. The coil configuration shown is suitable for determining diagonal components of this tensor. Off-diagonal components may be measured by means of primary coils of the types shown in figure 5.2(*b*).

A SQUID coupled to a gradient-sensitive primary is able to operate in magnetically noisy environments. Thus if the background field B is changing randomly with time but in a spatially uniform manner over the region occupied by the coils, no change will be detected. However a nearby 'localised' magnetic source will produce different field changes at the two coils, resulting in a net SQUID signal. Thus nearby signals can be detected in the presence of much larger distant fluctuations. SQUIDs have been used in this way to measure magnetic fields produced by the human body of order 1 pT in quite noisy magnetic environments where the ambient field may fluctuate rapidly with an amplitude of many nT (see sections 5.7 and 5.8).

5.3.1 *Higher-order gradiometers*

Even more complicated primary coils may be used to sense higher-order spatial derivatives of the field. To give some focus to a specific case, consider for example two identical first derivative pairs connected back to back. This arrangement will sense the second derivative of the field $\partial^2 B/\partial z^2$. Combining the two central coils which are wound in the same sense gives an effective geometry as shown in figure 5.2(*c*). The inductances of the various coils may be chosen to maximise the energy transfer to the SQUID arising from a particular field change. Suppose that, for the arrangement of figure 5.2(*c*), we wish to maximise the sensitivity to $\partial^2 B/\partial z^2$ but the space available for the flux transformer in the z direction must be less than z_0. In general for m coaxial coils spaced along the z axis at positions $z(i)$ with the ith coil having area $A(i)$ and $n(i)$ turns with a winding sense $W(i) = \pm 1$ the flux change $\delta\Phi$ produced in the primary by a field change $\delta B(z)$ is given by

$$\delta\Phi = \sum n(i)A(i)W(i)\delta B(i). \tag{5.6}$$

A Taylor expansion of δB about the point $z = 0$ gives (see figure 5.3)

$$\delta B(z) = [B_0 + (\partial B/\partial z)z + (\partial^2 B/\partial z^2)z^2/2 + \cdots].$$

The coil arrangement may be designed to measure a particular component of the gradient tensor, the jth say, $\partial^j B/\partial z^j$ by ensuring that all lower-order contributions to the gradient are zero, thus

$$\begin{array}{lll} \sum n(i)A(i)W(i)z(i)^0 & = 0 & \text{zeroth order} \\ \sum n(i)A(i)W(i)z(i)^1 & = 0 & \text{first order} \\ \sum n(i)A(i)W(i)z(i)^2 & = 0 & \text{second order} \\ \quad \cdots & & \quad \cdots \\ \sum n(i)A(i)W(i)z(i)^{j-1} & = 0 & (j-1) \text{ order} \end{array}$$

up to the $(j-1)$th term. This gives j equations and there are $3m$ unknowns for m coils (since $n(i)A(i)$ represents a single unknown under the assumption that the field $B(z)$ only varies in the z direction). Using the j conditions on lower-order terms being zero, together with boundary conditions relating to the allowed volume which the gradiometer can occupy and the optimisation of input energy transfer to the SQUID, it is possible in principle to design a circuit which responds to only a single term in the Taylor expansion of the field. In practice only third-order coils have so far been used.

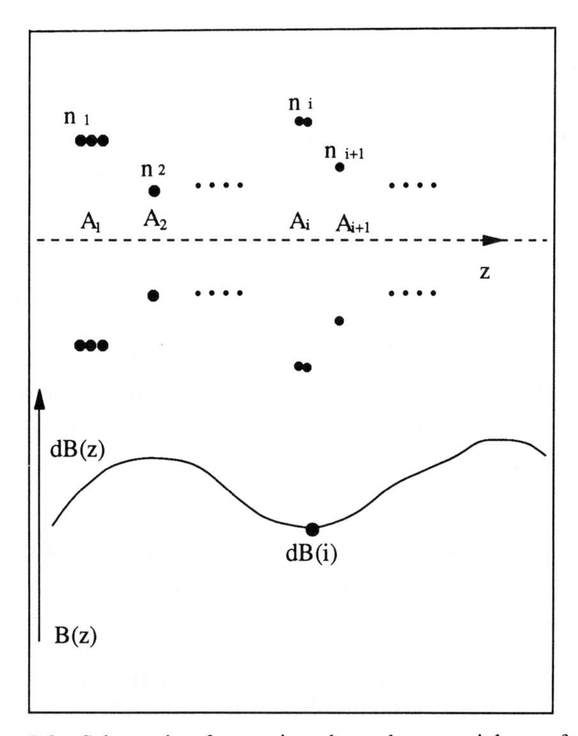

Figure 5.3 Schematic of a section through a coaxial set of coils

Thus it may be better to have those coils which are at either end of the transformer of unequal size in order to maximise the sensitivity. The science of gradiometer coil design has probably reached its apogee with the building of large coil arrays coupled to SQUIDs which serve as magnetic monopole telescopes. These have very complex two-dimensional sheets of coils carefully designed to minimise external influence while maximising the sensitivity and certainty of detection of the passage of a monopole (see Chapter 8).

5.3.2 *Multi-loop SQUIDs*

However ingenious the design of gradiometer coils, it is necessary to couple them as tightly as possible to the SQUID ring itself if the maximum available signal power is to be utilised. An interesting paradox arises in trying to achieve the most sensitive field detection within a given volume. It was pointed out in Chapter 4 that if intrinsic noise is dominant over other sources the minimum detectable input energy change in unit bandwidth is given by

$$\delta E = 8kTL/R \tag{5.7}$$

which depends directly on the loop inductance L. In general a small value of L requires a loop of small radius (although in principle the inductance of a long tube of fixed cross section is inversely proportional to its length and so may be made arbitrarily small). However the flux transformer volume should be as large as possible to maximise the available signal energy. It becomes increasingly difficult to couple a large primary coil to a very small SQUID ring via a secondary winding. Zimmerman (1972) pointed out an ingenious solution to this problem. If N superconducting rings, each of inductance L, are connected in parallel across the Josephson junction, the effective SQUID inductance will be L/N. One way to use such a multi-hole device is to sense magnetic field changes directly. Suppose that a cylindrical volume $V = \pi a^2 l$ is available and that we are always in the regime where the loop inductance determines the limiting sensitivity. Then the minimum detectable field change δB_1 for a single loop occupying the whole volume is

$$\delta B_1 = (\delta E/\mu_0 V) \propto (\pi a^2/lV)^{1/2}$$

whereas if the volume was filled with N loops of radius r and length l the minimum detectable change would be

$$\delta B_2 \propto (\pi r^2/lVN)^{1/2}.$$

However $N < \pi a^2/\pi r^2$ so that $\delta B_1/\delta B_2 = N$ and the field sensitivity just increases proportionally to the number of loops which can be connected in parallel. Zimmerman's scheme essentially involved bulk SQUID structures and was developed to a high level of sophistication in which the loops were given appropriate twists to allow different components of the field gradient tensor to be measured. The bulk devices have not been widely used since planar geometries are now favoured. However the multi-loop scheme lives on as a means of effectively coupling to very low-inductance planar DC SQUIDs (see section 4.4).

5.3.3 Gradiometer configurations for susceptibility measurements

The coil configuration shown in figure 5.2(*a*) is not limited to measuring field gradients. The whole arrangement may be enclosed in a superconducting shield to prevent it responding to any unwanted external magnetic source changes. Then if a material sample with some finite magnetisation is placed within one of the component coils, the total flux linking that coil will be changed. Even without inserting or moving the sample, the SQUID will show a change in output if the magnetic moment of the sample varies, for example with applied field or temperature (see figure 5.4). As an idealisation, suppose that the sample, of susceptibility χ, entirely fills one of the coils which has n turns of radius a. In an applied field B the additional flux linking the primary is

$$\delta\Phi = \pi a^2 n \chi B. \tag{5.8}$$

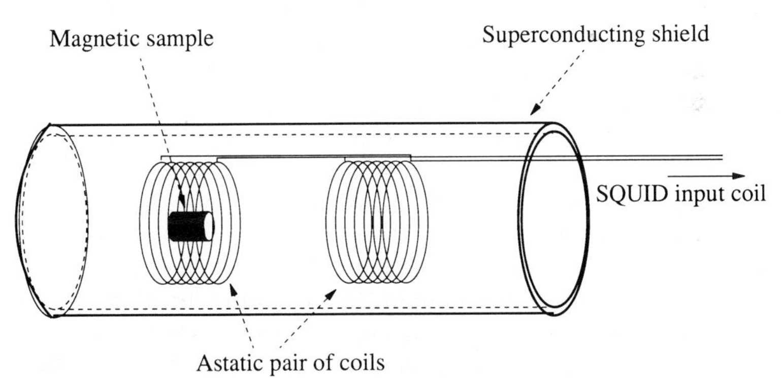

Figure 5.4 Schematic of a SQUID susceptibility transducer; the astatic coils are identical except that they are wound in opposite senses

For a SQUID with a flux sensitivity of $10^{-4}\Phi_0\ \mathrm{Hz}^{-1/2}$, typical of a commercially available system, with a primary coil of 100 turns of radius of 1 cm, the minimum change detectable in χB is of order 2×10^{-21} T in a 1 Hz bandwidth. To put this figure into perspective, it suggests that the nuclear paramagnetism of protons in water at room temperature and in the earth's field should be easily detectable. In reality the limiting sensitivity of the SQUID has never been approached in practical systems due to formidable problems with materials and constructional stability. Nevertheless SQUID magnetometers are far more sensitive than any other available type of instrument.

SQUID susceptometers of the general type described above have been used in a variety of applications. An early but impressive demonstration of the possibilities was made by Hirschkoff *et al* (1970) who measured the static susceptibility of copper nuclei in a field of 1 mT at a temperature of 10 mK. Such a system has been widely used as a nuclear susceptibility thermometer in ^{3}He–^{4}He dilution refrigerators.

Electronic magnetic moments being typically 10^3 times greater than nuclear moments means that tiny paramagnetic samples of this type may be measured. This sensitivity has been employed in a cerium magnesium nitrate susceptibility thermometer (Giffard *et al* 1972). Only a milligram or so of the salt is required, which means that the thermal resistance and heat capacity of the thermometer itself can be made very small. SQUIDs have also been used extensively in studies of the superfluid phases of ^{3}He. For a review of these topics see, for example, the article by Wheatley (1975).

Another type of susceptometer has been developed which has room temperature access to the SQUID pick-up coil, allowing rapid measurement of centimetre-size samples as they are passed through the primary. Dipole moments as small as 10^{-18} Wb m may be sensed (compare the magnetic moment of a single free electron $\mu_e = 10^{-24}$ Wb m). Susceptometers of this type have been used for quark searches in moon rock (Tassie 1965) as well as geophysical (Goree and Fuller 1976), archaeological (Aitken 1972) and biochemical (Cerdonio *et al* 1973) applications. The problems with utilising the intrinsic sensitivity of the SQUID are particularly severe when room temperature access is to be provided. Apart from the more obvious difficulties associated with ferromagnetic inclusions in supposedly non-magnetic construction materials, there are more subtle problems arising from thermoelectrically generated fields and dimensional changes associated with large temperature differentials within the apparatus, to which the SQUID becomes particularly sensitive if the applied field is also large.

Finally, a variant of the usual gradiometric susceptometer detects the change in static nuclear magnetisation of a sample in one coil when a resonance condition is satisfied between an applied RF field and the nuclear free precession frequency in the static applied magnetic field. This method of detecting nuclear magnetic resonance has been demonstrated for broad resonance lines at high temperatures (Day 1972) and for metals at low temperatures (Meredith *et al* 1973). Real-time detection of precessing nuclear spins has also been demonstrated. This is further discussed in section 5.10.

5.4 Electrical Measurements

To convert a SQUID to an ammeter or voltmeter is of course very straightforward. The application of a voltage or current to a coil inserted directly into the superconducting loop will directly change the magnetic flux applied to the ring. For the generalised electrical circuit of figure 5.5 the source will look like a voltage source if $R_s \ll R$ or a current source if the opposite limit applies. In this section we will consider only measurements at effectively zero frequency and also assume that the input coil is itself superconducting so that for all values of R the SQUID will see a current source. The change in applied flux due to a small change in voltage δV is easily seen to be

$$\delta\Phi = M\delta V/(R_s + R). \tag{5.9}$$

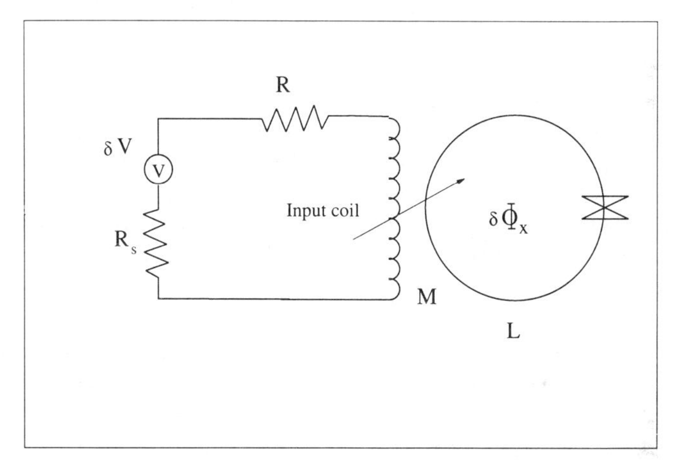

Figure 5.5 Generalised circuit diagram for electrical measurements using a SQUID

An estimate of the smallest detectable voltage change can be made by setting $\delta\Phi$ equal to the flux noise of the SQUID, taking as the limit of detection a signal-to-noise ratio of unity. For a low-impedance source and taking typical figures for commercial devices of $\delta\Phi = 1 \times 10^{-4}\Phi_0\ \mathrm{Hz}^{-1/2}$ and $M = 1\ \mu\mathrm{H}$, the voltage change detectable in a 1 Hz bandwidth becomes 10^{-19} V. This extraordinarily high sensitivity turns out to have found few applications at present. It is only for nano-ohm impedance sources at low temperatures that thermal noise approaches this voltage level, so for most normal measuring situations, in which there would be quite long normal leads connecting the source at room temperature to the SQUID, environmental noise is vastly greater than the limit set by noise in the SQUID itself. Considering the device as an ammeter, and using the same parameters,

$\delta i = \delta\Phi/M$ or around 0.1 pA (corresponding to about 10^6 electrons per second), at first sight a much less impressive figure than for the voltage sensitivity. However this sensitivity may be achieved with a room temperature source and has been used in a number of precision electrical measurements, including the cryogenic Josephson volt realisations and work on the quantised Hall effect as a resistance standard (Hartland 1981).

5.5 The Direct Current Comparator

An ingenious device which exemplifies the unique performance achievable with superconductivity is the direct current comparator, first proposed and realised by Harvey (1972). Consider a superconducting tube through which a wire passes, carrying direct current. This current generates a magnetic field, which, in order to screen the field from the interior of the walls of the tube, in turn induces a supercurrent to flow up the inside surface of the tube and back down the outer surface. Complete screening requires that the walls are at least several penetration depths thick, which is not difficult to achieve. The key point for the usefulness of this device is that the distribution of screening current on the outer surface of the tube is independent of the position of the current-carrying wire within the cross section of the tube, at least for all points on the outer surface which are more than a few tube diameters from either of its ends.

Suppose now that another wire is included in the tube which carries a variable current. The magnetic field produced by the supercurrent on the outer surface of the tube resulting from both wire currents can be measured with a SQUID, and zero field will be found when the two currents are exactly equal but opposite in direction. Because the external current distribution is extremely independent of the wire positions, and because SQUIDs are such sensitive devices for magnetic field detection, such a current comparator offers quite unparalleled precision for establishing a 1:1 current ratio, operating at any frequency from DC up to the maximum frequency of the SQUID input circuit. Unity ratios can be determined with accuracies of 1 part in 10^{12}, the sort of figure only previously associated with frequency comparisons.

5.5.1 Principle of operation

If we apply Ampère's law

$$\oint \boldsymbol{B} \cdot \mathrm{d}\boldsymbol{l} = \mu_0 i$$

to the three contours labelled a, b and c in figure 5.6(*a*) we find that

$$\oint_{a} \boldsymbol{B} \cdot \mathrm{d}\boldsymbol{l} = \mu_0 (i_1 + i_2)$$
$$\oint_{b} \boldsymbol{B} \cdot \mathrm{d}\boldsymbol{l} = 0 = \mu_0 (i_1 + i_2 + i) \tag{5.10}$$

where i is the current flowing on the inner surface of the tube. Since the field is everywhere zero on contour b, $i = -(i_1 + i_2)$. Invoking the requirement of current continuity we see that the outer surface current will just be equal and opposite to i:

$$i_s = -i = i_1 + i_2.$$

For a tube long compared to its diameter the current density distribution on the outer surface is obtained by minimising the self-energy of the resulting external field, quite independent of what goes on inside the tube. (This is just the inverse of the screening effect of the tube against external field changes, to be described in section 7.1.1.) A SQUID is then used to detect changes in the external field resulting from a change in current in either wire and the same change will produce the same effect regardless of which wire it occurs in. This transformer device can be extended by including n wires in series through the tube. This allows a current of exactly ni to be set up in the single winding, so the device is not limited to providing 1 : 1 ratios.

5.5.2 Practical current comparators

Usually the tube used to contain the windings is not straight but wrapped into a loop, its ends being overlapped, like a snake swallowing its tail. However the ends must not make electrical contact or the screening current flowing on the inner surface would not have to return over the outer tube surface. Such tubes are generally manufactured from lead foil. The transformer windings are made on a circular coil former which is then enclosed in a lead foil covering, the seams in the foil being joined by welding or soft soldering. The overlapped ends may be insulated from each other by mylar or adhesive tape. Great care is required when welding or soldering to prevent pin holes in the seams as these will seriously degrade the screening of the tube and thus reduce the accuracy of the current ratios. Figure 5.6 shows a schematic diagram of such a current comparator. The magnetic field changes produced externally by current changes are coupled into the SQUID by means of the usual flux transformer, the primary being similar in size and closely coupled to the tube itself. Of course any external field must be very stable and consequently the SQUID, comparator and

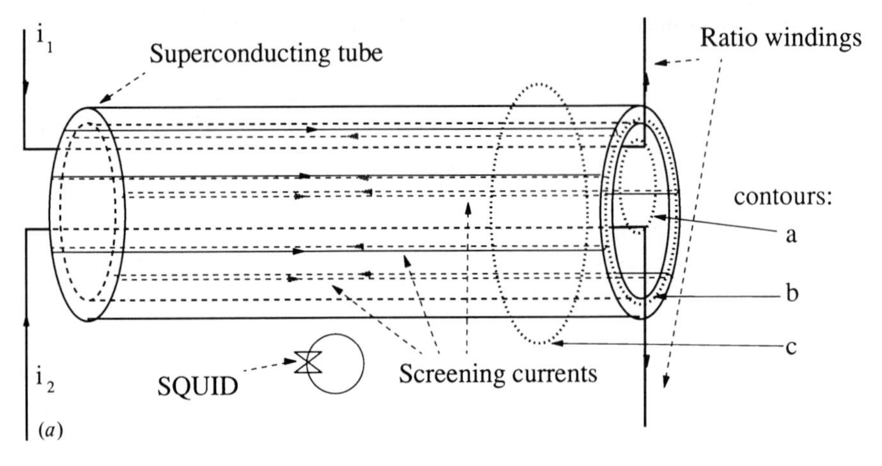

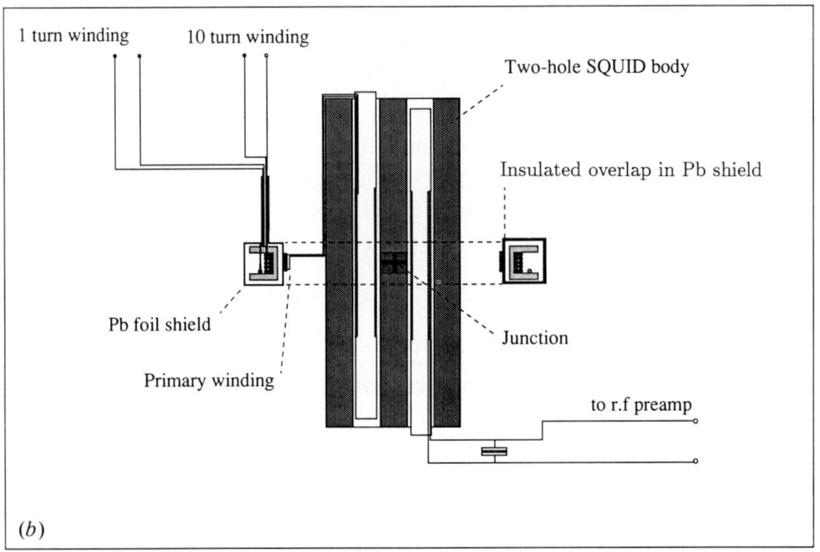

Figure 5.6 (*a*) Schematic of a direct current comparator; and (*b*) sectional view through a direct current comparator showing 10 : 1 ratio winding and two-hole RF SQUID

flux transformer must be extremely rigidly mounted in a well designed superconducting shield (see section 7.1).

5.5.3 Some applications of SQUID current comparators

A recent review of SQUID current comparators has been given by Gutmann *et al* (1989). To date most uses of these devices have been found in the field of precise electrical measurement. In work of this kind it is often necessary to calibrate a resistance with very high precision and the cryogenic current comparator is by far the most accurate method for doing this. The type of device outlined above has been widely used to provide resistor ratios in both cryogenic voltage standard systems (Hartland *et al* 1978) and in quantum Hall effect realisations of the standard of resistance (Hartland 1981). The ratio accuracy in these cases is typically a few parts in 10^9 for a current ratio of 400 : 1.

A portable secondary voltage standard has been developed by the author (Gallop 1977) using a 1 : 1 current comparator in a small (1 litre) liquid helium cryostat. This uses a closed loop of wire carrying a persistent supercurrent as a reference, to which an external current source may be stabilised using the comparator. The stable external current is then passed through a cryogenic resistor with a small temperature coefficient. The resulting voltage drop across the resistor is sufficiently constant in time to act as a reference which may be calibrated against the primary Josephson standard before being transported to other sites to calibrate local voltage standards. For a typical current of 100 mA and a 100 Ω resistor the calibration voltage can be at the 1 V level. Great care must be taken with shielding and rigidity if such a system is to withstand the rigours of transport. A prototype version has shown reproducibility to 0.1 μV.

5.6 SQUIDs for Alternating Current Applications

SQUIDs are not limited to measuring very slowly changing signals. In Chapter 3 the parametric amplification property of SQUIDs was outlined, showing (equation (3.15)) that the available power gain is

$$P = f_1/f_2$$

where f_1 is the pump frequency and f_2 is the signal frequency. In the case of a DC device the pump frequency is that of the internally circulating supercurrent generated by the voltage bias. One of the most significant developments in recent years has been the ability to produce devices which are pumped, either externally or internally, at a sufficiently high frequency so that adequate gain can be achieved at signal frequencies extending into the RF region.

In the early days of superconducting electronics the input bandwidths of SQUID devices were severely limited. Only single-junction devices were then generally available and these were typically pumped at a frequency of around 20 MHz. For this generation of SQUIDs the modulation frequency $\omega_1/2\pi$, used to linearise the response, was usually in the range from 10–100 kHz, limited by the requirements of adequate gain and readily available phase-sensitive detector circuit components. The input flux slewing rate is typically limited to $\sim\Phi_0/\tau$, where τ is the time constant of the integrator. Most simply $\tau > 2\pi/\omega_2$ so that the maximum slewing rate becomes $\omega_2/2\pi\Phi_0\ s^{-1}$. Thus it became accepted that only small signals in the audio-frequency band were capable of being measured. Beginning around 1976, various authors began work on single-junction devices pumped at UHF and even microwave frequencies. The anticipated increase in gain was reported but no widespread applications of the possible extension in signal bandwidth were forthcoming. It was only following the description by Long *et al* (1980) of a fully engineered RF SQUID pumped at 440 MHz, with a bandwidth of around 5 MHz and a flux sensitivity of $10^{-6}\Phi_0\ Hz^{-1/2}$, that the possibility of using a SQUID as a general purpose low-noise amplifier was seriously discussed. At about the same time developments in techniques for preparing tunnel junctions led to the possibility of producing DC SQUIDs with very small-area non-hysteretic tunnel junctions, with reasonably matched characteristics. As a result amplifiers based on thin film planar DC SQUIDs have recently been described with an even more impressive bandwidth capability. For example, Hilbert and Clarke (1983) have reported a noise temperature of 1 K for such a system and a bandwidth extending up to 200 MHz with a gain of 19 dB at 100 MHz. Their design incorporates a 20-turn Nb thin film input coil coupled to a planar DC SQUID, made up of 2 μm square $Nb-NbO_x-PbIn$ junctions. The significance of this work is that from DC up to around 1 THz, SQUIDs or other Josephson effect devices provide the lowest-noise detectors and amplifiers (see figure 8.9).

5.6.1 RF current measurements

Even before broad-bandwidth SQUIDs became available, an ingenious scheme had been developed (Frederick *et al* 1977) to measure the amplitude of an alternating current at radio frequencies. For a single-junction SQUID pumped at a microwave frequency the circuit may be analysed in terms of the amplitude and phase of the reflected microwave power from the superconducting ring. Both parameters will be a periodic function of the instantaneous flux applied to the ring. If this flux ($\Phi_1 \sin 2\pi f_1 t$) is varying with time at a frequency f_1 much less than the pump frequency but much greater than the upper limit of the detection circuit, the output will depend in a non-trivial way on the amplitude of the time-varying flux. If the output is $V(t)$ then

$$V(t) \propto \sin \Phi(t) = \sin(\Phi_1 \sin 2\pi f_1 t). \tag{5.11}$$

The SQUID system is unable to respond sufficiently rapidly to follow the instantaneous variation of flux at frequency f_1 but the time-averaged output $\langle V \rangle$ will vary with Φ_1, the signal amplitude. From the simple theory of frequency modulation it is readily shown that

$$\langle V \rangle = V_1 J_0(\Phi_1/\Phi_0) \tag{5.12}$$

where J_0 is the zeroth-order Bessel function, which is plotted in figure 5.7(*a*) to make clear its periodicity. Thus if flux Φ_1 is applied to the ring by a current i_1 flowing in a nearby circuit with mutual inductance M with the SQUID, the output will follow a Bessel function variation with the amplitude of this current and, for example, by counting zero crossings the amplitude may be deduced in terms of the (DC) current amplitude required to produce a change in applied flux of one flux quantum ($1\Phi_0$) at the SQUID. Such a device has been used at a number of national standards laboratories as a way of calibrating radio-frequency attenuators (Frederick *et al* 1977). A dynamic range of 105 dB and an uncertainty of 0.001 dB has been attained over a frequency range from DC to 1 GHz. Figure 5.7(*b*) shows a schematic of a waveguide microwave point-contact SQUID developed at NPL for RF calibration (Petley *et al* 1976).

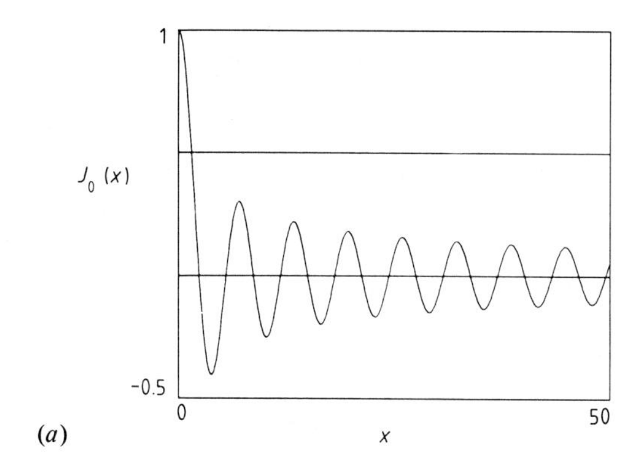

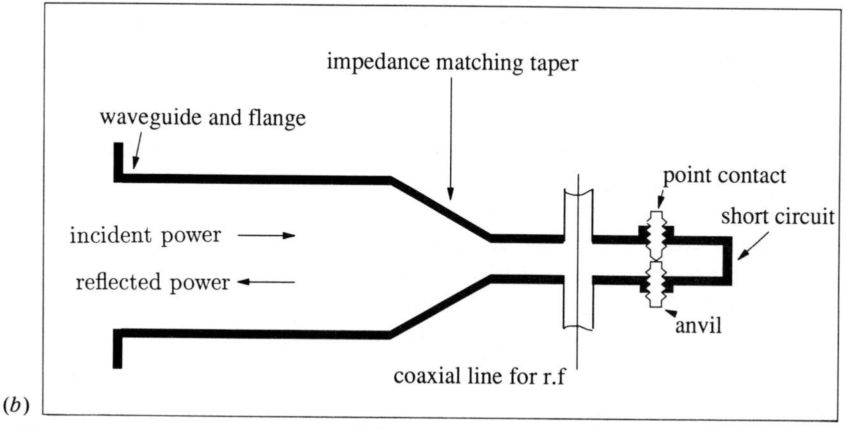

Figure 5.7 (*a*) Plot of the zeroth-order Bessel function $J_0(x)$, and (*b*) microwave SQUID mounted in a cast babbit superconducting waveguide (used for RF attenuator calibration)

The limitations with first-generation devices were set by small departures from a perfect sinusoidal current–phase relationship of the Josephson element. Tunnel junctions of small area and good uniformity of barrier height are now available which should be more accurately modelled by equation (2.8).

5.7 Geophysical Applications of SQUIDs

Having already pointed out that the extreme sensitivity of SQUIDs makes it impossible to operate them in an unshielded environment, it may seem perverse to claim that the devices can be used to measure the earth's magnetic field. To put this apparent contradiction in different terms: if fluctuations in the ambient magnetic field are so much greater than the limiting sensitivity of SQUIDs then why not use some conventional magnetometer with much less sensitivity? There is no single simple answer to this question, but in this section we shall summarise the main points of the argument.

First, it is interesting to compare the magnetic noise levels in a typical laboratory with those in a quiet open site, far from man-made electrical interference. The intrinsic fluctuations in the earth's field have typically a $1/f$ frequency dependence, with a magnitude of perhaps 10^{-11} T Hz^{-1} at about 1 Hz (although quite sharp peaks in the spectrum also occur in the range 0.1 to 10 Hz). In contrast, the magnetic noise in a laboratory may often have the same general frequency dependence but may be three orders of magnitude greater, with additional strong peaks in the spectrum associated with mains frequency, its harmonics and even sub-harmonics. In addition, ferromagnetic objects associated with the building structure itself or with apparatus, particularly if movable, will also be a significant source of low-frequency magnetic noise. Thus an unshielded SQUID which could not possibly remain flux-locked in a laboratory may behave quite well in open countryside.

But even if the SQUID remains flux-locked, why should its extreme sensitivity prove useful? A multi-turn induction coil is very sensitive to rapid changes in ambient field (Koch *et al* 1981). Campbell and Zimmerman (1975) compared a number of types of sensitive magnetometer for geophysical use and showed that not only was the high sensitivity of a SQUID useful, but also the compactness and relative simplicity of a SQUID system in a portable cryostat were a real advantage over the much more bulky (search coil) or fragile (optical pumping) magnetometers.

The aspect of geomagnetism in which SQUIDs have been most used so far is undoubtedly magnetotellurics. This technique involves the analysis of the response of the upper surface of the earth's crust to incident low-frequency electromagnetic disturbances propagating from the ionosphere. Due to the great height at which these disturbances originate they may be treated as plane waves. By simultaneously measuring the components of the magnetic field parallel to the surface and the resistivity of the surface layer itself it is possible, by analysing the frequency response of each, to deduce the impedance tensor of the earth's surface down to a depth of several kilometres. Typically the resistivity of earth ($1/\sigma$) ranges from $1\ \Omega\,\mathrm{m}$ to $10^4\ \Omega\,\mathrm{m}$ and the skin depth in km is given by the expression

$$\delta = (2/\mu_0 \pi f \sigma)^{1/2} \tag{5.13}$$

where f is the frequency in Hz. SQUID gradiometer sensitivities of $10^{-13}\ \mathrm{T\,m^{-1}}$ are easily attainable using commercial systems.

Measurements of this kind have been done for many years using very large search coils to measure the magnetic fields. Initially it did not seem that the increased sensitivity of a superconducting magnetometer would be worthwhile due to ambient noise. However by including two SQUIDs separated by several kilometres from one another, a distance great enough so that ambient noise at each is uncorrelated, it has proved possible to produce much cleaner data with a cryogenic system than with a conventional one. In addition, 'driven' magnetotellurics has been investigated with SQUIDs, in which a coil, carrying a large variable low-frequency current, is moved around while a fixed magnetometer senses the local field perturbations synchronous with the source frequency. A number of other proposals have been made for various magnetic surveying techniques associated with oil or mineral surveying, geothermal energy or earthquake prediction. A thorough review has been given by Clarke (1983).

Geophysical use of SQUIDs has provided a very important testing ground for real applications of superconducting devices. The environmental and logistical problems encountered are fairly extreme and the growing adoption of this technology by workers outside the security of a laboratory demonstrates the reliability and capability of such systems. A further significant impetus would probably come with the development of reliable, magnetically quiet, refrigerators for the liquid helium temperature range. Although portable dewars with a hold time between refills of up to two weeks have been built, it would be attractive to be able to run remote magnetometers continuously. There are a number of developments underway in this direction and they will be outlined in Chapter 7.

5.8 Biomagnetism

Until the development of SQUIDs, the detection of magnetic fields arising from biological functions was almost unknown as an area of study. In fact there exists a considerable number of sources of magnetic field associated with the human body. These range from ferromagnetic contamination and the susceptibility of tissue in an applied magnetic field, through ionic currents associated with healing processes to the rather weak magnetic signals resulting from muscular or neural activity. The order of magnitude of these effects goes from about 1 nT down to less than 1 pT. Figure 5.8 gives an indication of the strengths of various sources as well as the frequency range over which the signals are to be found. Comparing this figure with figure 5.9, showing estimates of noise levels for a variety of environments, it is clear that hardly any biomagnetic effects could be detected by a simple SQUID magnetometer in a typical laboratory environment. Two approaches to overcome this problem have been followed. First it is possible to enclose both SQUID and patient within a magnetically screened room. This is an effective but expensive solution. Ferromagnetic shielding of a given thickness is more effective if used in multiple layers. Up to six skins of mumetal have been used to give noise levels as low as 10^{-15} T $\mathrm{Hz}^{-1/2}$ (Mager 1982). A cheaper technique is to use thick non-magnetic metal shields to produce eddy current screening, which can be particularly effective at mains frequency (Zimmerman 1977). The second approach is to use a gradient-sensitive input coil configuration (see section 5.3). The principle relied on in this case is that noise sources are farther from the detector than is the signal. Typically the field gradient from any source falls off with distance faster than the field magnitude itself so that gradiometer coils are better able to discriminate against distant noise sources than is a single coil. Although a small sacrifice in sensitivity is involved, the ability to operate in an unshielded environment is a great advantage over the expense and immobility of a magnetically shielded room.

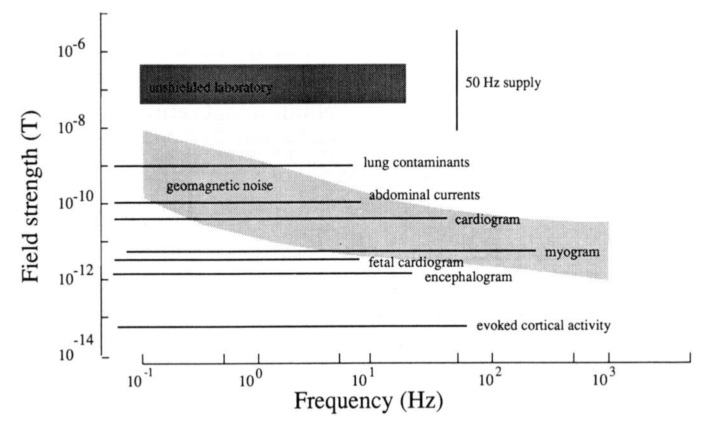

Figure 5.8 Typical amplitudes and frequency ranges of biomagnetic signals and some noise sources

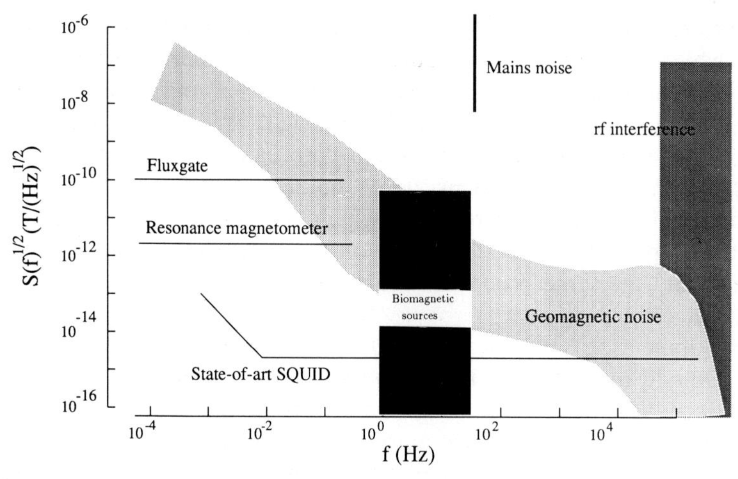

Figure 5.9 Spectral density of various magnetic sources and noise including sensitivities of some sensors

The relatively recent introduction of SQUIDs into medicine means that the diagnostic and clinical advantages of magnetic scanning of the body are still in the process of being proved useful. This field of study is at an early stage when the capabilities of the technique are being analysed and correlations established between clinical conditions and the signals received. It has already proved possible to make progress in a number of fields.

5.8.1 Neuromagnetism and evoked brain response

This is perhaps the area to which most work has so far been devoted. Brain activity produces magnetic field levels at the scalp of <1 pT and a SQUID is able to supplement information from electroencephalography (EEG) with its magnetic analogue (MEG). In the evoked response method one applies a repetitive stimulus, which may be visual, auditory or tactile, and signal-averages the SQUID output synchronously. The form of the response in amplitude and time should allow information to be deduced about the physiological processes going on in the brain as a result of the stimulus. The earliest efforts in this field involved moving the detector with respect to the skull. The relative amplitude of the magnetic field from the evoked response at a number of positions allows one to determine the spatial position and strength of the neural current source. To date the results mainly serve to confirm those already obtained from electrodes attached to the scalp. However, there is intrinsically more information available from remote field measurements than from electric potentials at the skull surface, which are inevitably modified by the intervening electrically conducting tissue and insulating skull between cortex and skin. The use of single-channel SQUID systems, which must be moved around the patient's head, limits severely the rate at which data can be collected. Ingenious, though rather bulky, systems with multiple SQUIDs and wire-wound gradient coil systems were reported which allowed real-time signal reception from up to seven points located on the surface of a sphere of radius 125 mm and the same group is working on a twenty-four channel system (Knuutila 1988).

Developments of planar DC SQUIDs with integral planar gradiometer coils (Donaldson and Bain 1984) hinted that arrays of detectors could be readily produced which would be arranged around a subject in a compact configuration to record a large amount of data in parallel, with a consequent increase in speed and reliability of results. Cross-talk between multichannel detector systems represents a particular problem which has yet to be addressed at the limit of SQUID sensitivity. Nevertheless, at the time of writing, such SQUID array systems with up to 37 separate channels are being tested in clinical applications, though few constructional details of the systems appear to have been published. A number of companies are offering such systems commercially at the time of writing.

5.8.2 Biomagnetism resulting from low-frequency currents

In addition to the alternating magnetic fields due to neural and muscular activity with characteristic frequencies around 1 to 10 Hz, there are also much more slowly varying fields originating in the human body. These are generated by quasi-DC currents, believed to be associated with organ development and regeneration. A relevant piece of work in this area, reported by Grimes *et al* (1983), presents a good example of the diversity of studies taking place in biomagnetism. Using a second-order SQUID gradiometer in a noisy unshielded environment they have scanned patients' bodies by placing them on a non-magnetic couch which moves in two orthogonal directions below the magnetometer. Large Helmholtz coils null the earth's field, and the noise present is less than 1 pT peak to peak in a bandwidth from 0.2 to 30 Hz. A survey of 14 apparently healthy pairs of human legs showed that typically a current of about 10 μA flows up one leg and down the other. The mechanism responsible for the current is not yet clear, although the source would appear to be in the muscles of the leg. This indicates a typical situation at present in biomagnetism. Good data (such as is shown in figure 5.8) is still awaiting physiological explanation. Another project carried out by the same group involves the examination of currents flowing in chicken eggs during development. In the early stages a simple field distribution is present which breaks up into more complicated distributions as development proceeds.

5.9 SQUID and Josephson Effect Noise Thermometry

In Chapter 4 the topic of thermal noise originating in a resistor R at temperature T has already been dealt with at some length. In the classical limit the voltage noise power per unit bandwidth is just

$$S(\omega) = V_n^2/R = 4kT.$$

In certain well shielded environments, this noise source, associated either with the resistance of the input coil or that of the Josephson junction, limits the sensitivity of SQUIDs. We will see in this section that it is possible to put to good use this otherwise troublesome effect, since the noise power provides a means of measuring the absolute temperature. This is not a new idea, for high-temperature noise thermometers have been known since 1945, but it is only with the advent of SQUIDs (with their extraordinary sensitivity) that this type of measurement of temperatures has developed, not just in the liquid helium range but, as we shall see, down to the millikelvin region.

Two types of Josephson effect noise thermometer have been described, the first of which simply uses a SQUID as a very sensitive preamplifier in a conventional circuit. The second incorporates a resistor in a SQUID ring, allowing the Josephson junction to be biased by a direct current into the finite voltage regime. Consequently a Josephson supercurrent will flow, with a frequency defined by the bias voltage. The Johnson noise voltage fluctuations will frequency modulate this Josephson 'carrier frequency'. A determination of the linewidth of this radiation allows the absolute temperature T to be deduced. Both types will be discussed in the following sections, describing the work on each which has been reported, and considering future designs using newer lower-noise SQUIDs, with emphasis on how these thermometers might be 'speeded up'. A final paragraph considers a type of DC SQUID as another possible device configuration.

5.9.1 Conventional noise thermometry with a SQUID preamplifier

A major problem with high-temperature noise thermometers using conventional electronics is that it is necessary to ensure that the frequency response (and thus an effective bandwidth) of the instrument should be well defined and independent of both temperature and time. If the bandwidth cannot be accurately characterised, the thermometer must be calibrated at a known temperature and is not therefore an absolute instrument. Nevertheless the existence of superconducting transition reference thermometers has provided reliable fixed temperature points, so that the linearity of the noise thermometer provides an accurate interpolation instrument. Consider the circuit shown in figure 5.10. It is assumed that the defining bandwidth is set by the characteristics of the input circuit, thus

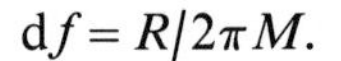

$$\mathrm{d}f = R/2\pi M.$$

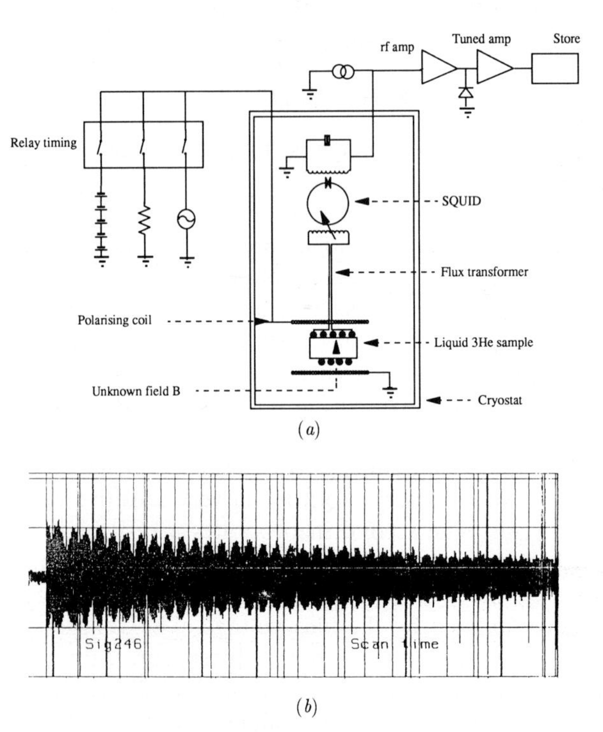

Figure 5.10 (*a*) Schematic of HE-SQUID magnetometer; and (*b*) sequence of 40 spin echos lasting 80 s

Then the mean square current noise is

$$\langle i^2 \rangle = 4kT/2\pi M. \qquad (5.14)$$

The SQUID output is sampled at time intervals $\mathrm{d}t$ where $\mathrm{d}t \gg 1/\mathrm{d}f$ and the mean square value of the output calculated. The SQUID of course has intrinsic noise which limits the ultimate sensitivity achievable in a given measuring time. Assuming that the minimum detectable current change δi is given by the minimum detectable energy change formula (equation (4.15))

$$\delta i^2 M = \delta E = \delta A \, \delta f = 2kT_{\mathrm{n}}$$

where T_n is the noise temperature of the SQUID. To achieve a temperature measurement of precision $\delta T \sim T_n$ at a temperature T will require a total time t where

$$t > (kT)^2/(\delta A^2 \, df^3). \tag{5.15}$$

To take an example, the energy sensitivity of a commercially available RF SQUID operating at 20 MHz is about 10^{-28} J Hz^{-1}. Thus to achieve a measurement of 1 mK precision at a temperature of 100 mK will take about 100 s. This type of thermometer has a number of distinct advantages. For example the power input to the low-temperature region can be made vanishingly small, no external power being dissipated in the resistor. The resistor can be remote from the SQUID preamplifier, being connected by low-inductance superconducting leads, which at very low temperatures will have the added advantage of extremely low thermal conductivity. Thermal contact to a metallic resistor is also much better than that to other materials such as insulating crystals or liquid helium which represent possible alternative thermometric substances. The relatively slow measurement process is also not a serious problem in the temperature range below 1 K since thermal contact becomes increasingly difficult here, leading to slow temperature equilibration times and prolonged experiments.

5.9.2 *Experimental results*

Webb, Giffard and Wheatley (1973) have described a conventional SQUID noise thermometer in considerable detail, considering its limitations theoretically, and comparing it experimentally with a magnetic susceptibility temperature scale based on cerium magnesium nitrate (CMN). This group has reported agreement to within 1% between the noise thermometer and the susceptibility thermometer. At the lowest temperatures reached, about 10 mK, the two instruments showed a significant discrepancy of about 2 mK. The authors attribute this to a loss of temperature equilibrium between the two thermometers, although it has also been suggested that departures from the classical noise processes in metallic resistors occur at these temperatures and that such effects are responsible for the discrepancy.

5.9.3 *Josephson linewidth thermometer*

As described above, the junction is biased into the finite voltage region, when the noise voltages frequency modulate the Josephson alternating supercurrent. Voltage noise in metallic samples is believed to be white up to very high frequencies, comparable with the lowest normal resonant mode of the specimen, but the effect of a highly non-linear device like a Josephson junction is to compress the noise power spectrum into a well defined bandwidth around the carrier frequency. Kamper and Zimmerman (1971) realised the importance of the fact that the bandwidth is itself directly proportional to the absolute temperature, the constant of proportionality being a fundamental physical constant. It is this feature of the thermometer which is particularly attractive since the bandwidth does not need to be measured, nor is a calibration point required for this absolute instrument.

5.9.4 Bandwidth compression

A simple argument demonstrates the principle of bandwidth compression and its importance for this type of noise thermometer. Consider a Josephson junction shunted by a small resistor R, through which a constant current i flows, biasing the junction with a time-averaged voltage V_0. Then at $T=0$ K the junction current would oscillate at a precise frequency $\omega_0 = 2\pi V_0/\Phi_0$. At a finite temperature T imagine a noise voltage component $V_n(\omega)$ across the Josephson junction at frequency ω. The effect of this component on the Josephson oscillation is to frequency modulate it, giving an output of the form:

$$A \sin(\omega_0 t + (\Delta\omega/\omega) \sin \omega t)$$

where A is a constant. For a Josephson junction, $\Delta\omega = 2\pi V_n/\Phi_0$ from the basic AC Josephson effect relationship. Thus the modulation depth is

$$m(\omega) = \Delta\omega/\omega = 2\pi V_n(\omega\Phi_0)^{-1}.$$

Qualitatively we can say that the output at frequency $\omega_0 + \omega$ is only significant if $m(\omega) > 1$. The frequency difference ω between ω_1, for which $m(\omega_1) = 1$, and ω_0 can be taken as half the effective linewidth $\mathrm{d}f$ arising from this suppression:

$$\mathrm{d}f = 2\omega/2\pi = 2V/\Phi_0 = (2/\Phi_0) \times (4kTR\ \mathrm{d}f)^{1/2}$$

(remember, the noise-voltage frequency modulates on both sides of ω_0). Note that $\mathrm{d}f$ appears on both sides of this expression since, on the right-hand side, the noise power is directly proportional to it. Simplifying this expression gives a result for $\mathrm{d}f = 16kTR/\Phi_0^2$. This rough argument gives an answer which is $4/\pi$ greater than the exact result. Thus it is the fundamental voltage-to-frequency conversion constant of the junction which means that the device imposes its own temperature proportional bandwidth on the thermal fluctuation spectrum. It is of course necessary, if this bandwidth is to be accurately measured, that the bandwidth of subsequent amplification stages should be much wider.

There are further points to consider in the noise thermometry field. No serious work has appeared to the author's knowledge in which a DC SQUID, with unshunted junctions, is used directly as a noise thermometer. Such a system, having very high gain and a very high pump frequency, would be very much faster than the Josephson noise thermometers so far reported. For this hypothetical thermometer the noise-generating element appears in the circuit after amplification, as well as in the input circuit, and this would certainly complicate analysis considerably. Screening of the noise-generating resistor and the SQUID itself becomes more and more important as the temperature goes down. A toroidal geometry would appear to be ideal.

It would also be possible in principle to measure the thermal noise rounding of the $I-V$ characteristic of a DC SQUID (or even of a single junction), since the rounding is ideally a simple function of temperature. This should provide a very fast system, but great attention would have to be paid to the shielding of the Josephson device and the associated circuitry.

5.10 An Absolute SQUID Magnetometer

Although very small absolute magnetic flux changes may be readily detected with a SQUID, the device must be calibrated if required to determine flux density changes. An alternative scheme has been developed which provides a measure of both the absolute value of, and changes in, the flux density without calibration, and which under favourable conditions can approach the sensitivity of a conventional SQUID magnetometer. The device is a cryogenic analogue of a proton resonance magnetometer. The water sample is replaced by liquid ^{3}He at a temperature of 3 K or below and a SQUID is used to detect the freely precessing nuclear magnetisation, rather than detecting an induced voltage with a low-noise semiconductor amplifier.

Consider a pick-up coil, the primary of a flux transformer coupled to a SQUID, with volume V and length l. Imagine that a single nucleus with spin I and magnetic moment μ_n is situated inside the coil. The magnetic moment will produce a change in the total flux linking the coil so that if the spin were aligned along the coil axis and its direction then reversed there would be a corresponding change in the flux linking the coil which in principle might be detectable. The coupling of a single nuclear spin to the coil will be position-dependent, but we may simplify the situation by assuming that the nucleus moves rapidly throughout the interior of the coil, averaging out the position-dependent coupling. Hence the magnetisation M, that is the magnetic moment per unit volume, is seen to be

$$M = \mu_n / V \qquad (5.16)$$

and the additional flux contribution through the coil $\delta\Phi$ is

$$\delta\Phi = \boldsymbol{B}\cdot\boldsymbol{A} = \mu_0\mu_n A/V = \mu_0\mu_n/l \qquad (5.17)$$

whereas the energy change on spin-flip coupled to the SQUID becomes

$$\delta E = (\delta\Phi)^2/2L = \mu_0\mu_n{}^2/2Vn^2 \qquad (5.18)$$

where n is the number of turns on the coil, whose inductance is L, and it has been assumed that the long-solenoid inductance approximation may be made. For a coil with a volume of 1 mm^3, even with a single-turn coil, $\delta\Phi \sim 10^{-29}$ Wb and $\delta E \sim 10^{-49}$ J so that the magnetic moment of a single nucleus in such a coil is very far from being detectable with any SQUID now available. Setting a limiting energy resolution of $\hbar/\mathrm{d}f \sim 10^{-34}$ J Hz^{-1}, which is approached by the best low-noise SQUIDs, we may ask how many nuclear spins would be required to be detectable. For a typical nuclear spin the number δN is of order 10^3 for the coil dimensions given above and a bandwidth of 1 Hz. Clearly this is the required excess of spins of one alignment over the number with the opposite. In practice such an excess spin population might be achieved by one of two basic methods, which will be dealt with in the following two sections.

5.10.1 Brute force polarisation

Application of a magnetic field causes the Zeeman splitting of up and down spin states for a spin $I = \frac{1}{2}$ nucleus. If a system of spins is in thermal equilibrium with a bath at temperature T the differential population of spin states is given by the Boltzmann formula, giving rise to a net nuclear polarisation. Due to the small size of nuclear moments the polarisation produced by this 'brute force' method will be small for reasonable laboratory fields in the kelvin temperature range. The difference in up and down spin populations may be calculated from the expression linking M with the susceptibility χ given by the usual Curie law formula:

$$\chi = N\mu_0\mu_n{}^2/3kT \qquad (5.19)$$

so that

$$M = \chi B/\mu_0 = N\mu_n B/3kT \qquad (5.20)$$

where B is the polarising field and T is the temperature of the spins. For a typical nuclear spin at 1 K a field of around 1 T is required to produce a fractional population difference of 1 in 10^4. It is interesting to note that the simple theory used above to determine the available energy per nuclear spin shows that this varies inversely with volume, whereas for a fixed spin density the number of spins is proportional to V. Thus the available signal should be size-independent, provided that the sample volume can be correctly matched to the SQUID, but the higher the density of polarised spins, the better. For the above example we need a density of 10^{15} spins/mm^3 to be detectable and using brute force polarisation requires an overall spin density of 10^{19} mm^{-3}. This corresponds roughly to the density of gaseous ^{3}He at a temperature of 1 K and a pressure of 100 mbar.

5.10.2 Optical pumping

As an alternative it is also possible to produce very high fractional nuclear spin polarisations in He by the technique of optical pumping. He gas is irradiated with circularly polarised light from a ^{4}He source (a laser tuned to the $S_{1/2}$–$P_{1/2}$ produces the highest degree of polarisation). The 1.08 μm radiation promotes a preferential population of the metastable 2^3S_1 states which in turn produce polarisation of the Zeeman split ground states by metastability exchange transfer, allowing up to 70% polarisation to be achieved if a laser is used for the optical pumping. With appropriate phasing of the applied RF pulses a maser action may be produced, allowing CW detection of free precession (Richards *et al* 1988). This seems obviously better than the brute force method but has a limitation. Optical pumping can only occur in He gas if the pressure is less than about 1 mbar. Higher pressures result in rapid depopulation of the metastable state by collisions. Thus the maximum polarised spin density which may be produced by this method is around 10^{19} mm^{-3}, strictly comparable with the density which may be produced in the liquid He of much higher density by a reasonable brute force field. Thus neither method is at present clearly superior.

5.10.3 Experimental systems

Both polarisation methods have been used (Taber 1974, Gallop and Radcliffe 1981) in absolute ^{3}He magnetometers using SQUIDs which we name HE-SQUIDs. It is interesting to calculate the sensitivity of such a magnetometer compared with that of a conventional magnetometer using the same SQUID, and having the same available sensing volume. The minimum detectable field change with the HE-SQUID is simply

$$\delta B = \delta\omega/\gamma = \Phi_n/\mu_0\beta\,|\,M_0\,|\,\gamma t \tag{5.21}$$

where γ is the ^{3}He gyromagnetic ratio, β is the flux-to-field transfer function, which depends only on the volume and SQUID parameters, and t is the time available for a measurement. Φ_n is the equivalent flux noise of the SQUID. The minimum detectable field change with a simple SQUID magnetometer is

$$\delta B = \Phi_n / \beta \tag{5.22}$$

where the parameters have the same meaning as above.

The flux signal appearing at the SQUID input from a sample of freely precessing spins with total magnetisation M_0 is

$$\Phi_s = \mu_0 \beta M_0 \sin \omega t.$$

In a brute force polarising field B_p the magnetisation M is given by

$$M_0 = \chi B_p / \mu_0 \tag{5.23}$$

where χ is the nuclear susceptibility. Thus the ratio R of the minimum detectable change in B for the free precession magnetometer compared to that of the simple SQUID becomes

$$R = 1/(\gamma \chi B_p t). \tag{5.24}$$

For liquid ^{3}He at a temperature of 3 K the susceptibility $\chi \sim 10^{-7}$ so that $R < 1$ for $B_p > 0.1$ T and measuring time $t > 1$ s, showing that under these conditions the free precession magnetometer is more sensitive.

One difference between this cryogenic NMR magnetometer and a proton free precession device involves the field dependence of the signal. Since the SQUID detects flux, the magnitude of which depends on B_p and temperature but not on B itself, the sensitivity is field-independent. The proton device detects an induced voltage signal derived from the rate of change of flux, so that its sensitivity increases with field B. Below about 10 μT these devices stop functioning, whereas the HE-SQUID will operate, at least in principle, in zero field without loss of gain.

Another important noise consideration for the HE-SQUID involves $1/f$ noise processes. We have seen elsewhere that these dominate white noise in SQUIDs below some frequency in the range 0.1 Hz to 1 kHz, which depends on the construction of the device and the nature of the Josephson junction. If the HE-SQUID can be operated in a small stable ambient field B_{ref}, such as may be trapped in a well designed superconducting shield, then low-frequency signals appear as sidebands on the basic precesssion frequency $\omega_0 = \gamma B_{ref}$, which can be arranged to be within the white-noise-dominated part of the SQUID spectrum. Then it should be practicable to carry out demodulation after several stages of gain without introducing significant $1/f$ noise.

5.11 SQUIDs for Conventional NMR

The extended discussion above is concerned with audio-frequency nuclear free precession experiments, currently a rather narrow field of application where SQUID systems have a unique advantage over conventional ones. A very interesting recent development is a broad-band ultra-low-noise RF amplifier based on a high-quality thin film design (Hilbert and Clarke 1983). A 20-turn planar signal coil with an input inductance of 0.12 μH is coupled to a DC SQUID made from two Nb–NbO_x–Pb(In) junctions (see section 6.2.1), shunted by 7 Ω resistors which provide adequate damping so that hysteresis is avoided. The demonstrated gain is 20 dB, approximately constant over the frequency range from DC to 160 MHz, and at an operating temperature of 1.5 K the noise temperature of this system is ~ 1 K, some two orders of magnitude lower than that of the best room temperature RF amplifiers. The importance of this development is that an amplifier of this kind may rapidly find general use in NMR systems where the very best low-noise performance is required. Since many high-field, high-homogeneity NMR spectrometers already employ liquid-helium-cooled superconducting magnets, there should be relatively little consumer resistance to the idea of cryogenic amplifiers.

The simple analysis of sensitivity of NMR SQUID systems given in the previous section begs a number of questions concerning noise processes in NMR systems. This topic does not seem to have been thoroughly discussed in the literature, where the assumption is widespread that the dominant source will be Johnson noise in the pick-up coil or following amplifier. Whereas this is well justified for conventional room temperature coils and preamplifiers, the assumption needs to be re-examined for superconducting coils and SQUID detectors. Unfortunately the intrinsic thermal noise in a system of polarised spins is difficult to quantify since inevitably the system is far from thermal equilibrium and a well defined spin temperature may not exist. A simple analysis, due to Gallop and Radcliffe (1981) suggests that the equivalent flux noise is proportional to the square root of the ratio of the measurement time t to the spin-lattice relaxation time T_1. Due to the very large value of T_1 for ^{3}He at both ambient and cryogenic temperatures, the nuclear spin thermal noise source seems to be negligible under almost all circumstances, even for the lowest-noise SQUID system. Recently Hilbert and Clarke (1983), using a nuclear quadrupole resonance system based on the SQUID amplifier described above, have observed this nuclear spin fluctuation noise, which is in good agreement with their theoretical predictions.

6

High-frequency and Digital Applications of the Josephson Effects

We have already seen, in Chapter 2, that the complex response of a Josephson junction to applied electromagnetic radiation forms a basis for both broad-band and heterodyne detection modes, ranging from low frequencies to the far-infrared. Greater detail of actual working devices of this type will be given in this chapter. The second topic dealt with is digital applications of SQUIDs. This is already a significant, and may in the future become a dominant, use of low-power superconducting technology. At first sight the contents of this chapter may seem to divide into two rather disconnected areas. However, logic devices have two important characteristics in common with radiation detectors: first, both involve very high-frequency operation, since slow digital devices are of little or no practical interest. Second, both applications demand the use of well characterised and reproducible small-area junctions with similar parameters. In what follows, more details of junction fabrication methods will be given than have been discussed earlier in the book. This reflects the fact that well defined junction properties are of the utmost importance in achieving high performance in these areas. It seems paradoxical that SQUIDs, for all their unparalleled sensitivity, will operate very satisfactorily with far from ideal Josephson junctions. This is mainly because SQUIDs have only been used in rather low-frequency applications, for which the low resistive and reactive impedances associated with poor-quality junctions do not limit performance.

6.1 Requirements for High-quality Junctions

For logic applications there are a number of desirable features which

Josephson junctions should possess. Of prime importance is small size since the main advantage of the superconducting environment is low power dissipation which allows very high packing density. This property can only be exploited if the device size can be adequately reduced. Almost equally desirable is that small size and compact interconnections minimise the propagation delay between components. Already in semiconducting devices these delays represent a significant contribution to the limits on computing speed. Small size also ensures that both the junction normal-state conductance ($1/R$) and its shunt capacitance C will be small. The significance of this is that a large value for R allows effective impedance matching to external transmission lines, while the quantity $1/RC$ sets one of the limits to high-frequency operation. For a tunnel junction biased below the gap voltage the quasi-particle leakage current should also be small so that the current change on switching is as large as possible.

The requirements are identical in the case of junctions for detection of electromagnetic radiation. Small conductance and capacitance ensure that high-frequency currents flow through the non-linear component of the junction, rather than through parallel, ohmic paths which contribute nothing to the mixing and detection functions. Additionally in this case the overall size of the junction and its connections must be much less than the free space wavelength of the incident radiation, a condition which is usually easily satisfied. In subsequent sections we describe some general fabrication techniques for thin film junctions of various types, ending with a little more detail of one specific process under development in Japan, which employs only refractory materials, giving a number of advantages.

6.2 Tunnel Junction Fabrication

Three main aspects of junction fabrication will be outlined in the next three sections: (i) electrode materials and deposition, (ii) pattern definition and (iii) barrier preparation. In any real fabrication process, particularly when a large number of interconnecting devices are required, there are many additional processes, including, for example, planarisation, insulation, interconnection and groundplaning. A description of these is beyond the scope of this book but details may be found in the references set out in subsequent sections.

6.2.1 Electrode materials

Although junctions have been made using electrodes of almost all elemental superconductors, there are really only two materials, lead and niobium, which have enjoyed widespread use. These two metals have the highest elemental transition temperatures (7.2 and 9.2 K respectively). Their material properties are very different. Lead is a soft, low melting point, metal which is easily oxidised to a considerable depth. Niobium melts at 2500 K and is highly reactive, forming in air a thin, tough oxide Nb_2O_5 which prevents further oxidation. Lead films kept in air at room temperature rapidly oxidise right through, destroying superconductivity at low temperatures. Pure lead electrodes are unsuitable for junctions which must be cycled to room temperature because the thermal expansion mismatch between lead film and substrate causes buckling and fracture of the film surface, which in turn leads to disruption of the insulating tunnel barrier. Alloying the lead with various metals produces much more stable films. If these are fine-grained (achievable by evaporation onto a substrate cooled to 77 K), their ability to survive thermal cycling is still further improved. A lead alloy process extensively developed by IBM consisted of a base electrode of Pb–12%In–4%Au and a counter electrode of e phase Pb–31%Bi (Huang and Schad 1983), on a substrate of high-purity silicon.

In addition to pure Nb electrodes, a number of Nb alloys have been explored, in view of their relatively high T_c's. The metallurgy of these A-15 compounds is complex and problematic, as is the formation of robust, continuous insulating oxides on them. However further progress in this direction may be expected. Perhaps surprisingly, the compound NbN is superconducting with a transition temperature in the bulk form as high as 15 K. By sputtering Nb through a nitrogen atmosphere it has proved possible to deposit thin films of the material which have T_c's as high as 14.5 K, and these have been used to make some of the best tunnel junction characteristics which have been demonstrated.

Recently much effort has been directed at producing thin film Josephson junctions using copper oxide ternary compounds. At the time of writing the 'workhorse' compound, $YBa_2Cu_3O_7$, has a high reproducible transition temperature (93 K in the bulk state) and has been produced in the superconducting state in thin film form by a number of routes. Although no clearly preferred fabrication techniques have arisen for the ceramic superconductor, brief details of some of the methods so far developed will be included where appropriate in the following sections.

6.2.2 Electrode deposition

The metallic planar electrodes, between which the tunnelling structure is formed, should be at least 100 nm in thickness, to exceed both typical penetration depth and coherence length of the material so that bulk superconducting properties are approached. For low melting point materials (below about 1500 °C), evaporation in high vacuum may be achieved by simple resistive heating in a refractory crucible. To prevent thin film contamination the pressure should be of order 1 mPa or better and the film deposition rate at least 1 nm s^{-1}. High-temperature copper-oxide-based superconductors have been evaporated by a simple ablation process in which a pulsed excimer laser is directed onto a bulk sample of the ceramic. The high input of energy in a very short time ensures that the material is evaporated in 'lumps' consisting presumably of many unit cells of the structure so that the deposited film has a composition very close to the starting material. The resulting material must be either carefully annealed in an atmosphere of oxygen at a temperature of >850 °C before it becomes superconducting or else deposition must be carried out onto a substrate heated to around 650 °C in the presence of an oxygen flow. This latter *in situ* process produces some of the best available films with T_c's for the YBCO ceramic on an MgO substrate as high as 91 K and with critical current densities $>5\times10^6$ A cm^{-2} at 77 K (Venkatesan *et al* 1989).

In the case of refractory metals, such as Nb, other methods must be used. Most common is sputter deposition, in which an intense RF or DC field produces an energetic plasma in an inert gas atmosphere in the region between a target electrode made of the material to be evaporated and the substrate, causing emission of metallic ions with subsequent deposition on the substrate (see figure 6.1). For evaporation the gas is usually chemically inert (e.g. argon) at a pressure of around 0.1 Pa. For film etching, carbon tetrafluoride (CF_4) or molecular oxygen (O_2) is admitted to the chamber. Good superconducting niobium films with transition temperatures as high as 9.1 K have been produced in this way. Although the technology is expensive, much of the equipment is available commercially as a result of semiconductor industry requirements.

Another method which has produced good results is evaporation by electron beam bombardment. A carefully focused high-power electron beam is directed at the material to be evaporated, which is placed on a water-cooled hearth. The electron beam power must be around 1 kW to produce reasonable deposition rates. Both sputtering and electron beam evaporation have been used for ceramic superconductors, preferably employing multiple sputter targets or electron beam sources, one for each metallic component in the final compound. Transition temperatures have been achieved for $YBa_2Cu_3O_7$ which are very close to the best bulk values. Again post deposition annealing in an oxygen atmosphere is required unless *in situ* deposition onto a hot substrate in a low-pressure local oxygen atmosphere is possible.

Various starting materials have been used, but oxidation of barium metal has led to problems so that the fluoride is used as the starting material, resulting in a significant improvement in resistance to corrosion and atmospheric degradation, and very high critical current densities (3×10^6 A cm^{-2} at 81 K).

Very high vacuum is required when depositing refractory pure metals since the raised temperature means that evaporated atoms react more readily with impurity atoms and also their high kinetic energy means that impurities may become embedded in the deposited metal film.

The substrates used are generally of silicon, quartz or glass, although sapphire has a number of low-temperature advantages which may lead to increased use in the future, in spite of its higher cost. Silicon is particularly favoured in that it is available cheaply in large, high-purity monocrystalline slices (again thanks to the semiconductor industry) and the technology for dicing wafers, attaching leads etc is also readily purchased. Special substrates give better performance with copper oxide materials. $SrTiO_3$ is particularly popular, since although expensive it has a similar perovskite structure and lattice spacing to the high-T_c materials so that with careful deposition and annealing, almost perfect single crystal films may be deposited epitaxially. MgO and $LaAlO_3$ also allow epitaxial growth and have superior properties to $SrTiO_3$ for microwave applications.

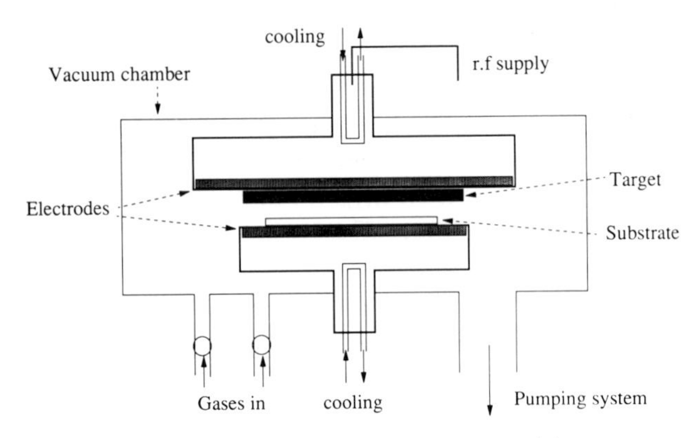

Figure 6.1 Basic elements of a sputter deposition system

6.2.3 Patterning of thin films

The simplest method for patterning thin films uses metal masks with slots cut in the shape of the film to be deposited, which are placed in good mechanical contact with the substrate and exposed to an incident flux of the evaporated material. This works well provided very small features or very accurate edge definition are not required. The limiting resolution is of the order of 25 μm. Multi-layer structures have been produced, using complex mechanical mask-changers which permit successive evaporations, possibly using several different materials, to be performed without breaking the high vacuum. The masks must be thin in order to reduce the effects of 'shadowing' produced at the edges, arising from the finite solid angle subtended by the source. Stainless steel is an excellent mask material, providing strength in thin sheets. Other more elaborate mechanical mask techniques have been used in which the mask is thick enough to give good rigidity but is thinned down only around the cut-out slots which define the pattern. Even with extremely good contact between mask and substrate some migration of atoms along the surface following deposition is inevitable and it is this process which limits the resolution of mechanically masked patterns.

6.2.3.1 Optical photolithography

For features with a resolution down to 1 μm a different technique is required and optical lithography proves suitable. In outline this involves depositing a uniform layer of metal which is patterned using a photosensitive layer, optically exposed and then developed. This is followed by a process to remove metal from unwanted areas. Many variations on this theme exist. A commonly used one is the 'lift-off' process. Initially a photosensitive 'resist' layer is spun on, as a uniform coating over the substrate, which is then exposed through a contact negative print of the required pattern, using ultraviolet light. The pattern is developed, removing the exposed portions of the layer. Subsequently metal is evaporated over the whole substrate which, on immersion in a solvent, leads to removal of the remaining photosensitive layer, taking with it the metal deposited on top of it while metal evaporated onto the bare substrate remains undisturbed (see figure 6.2).

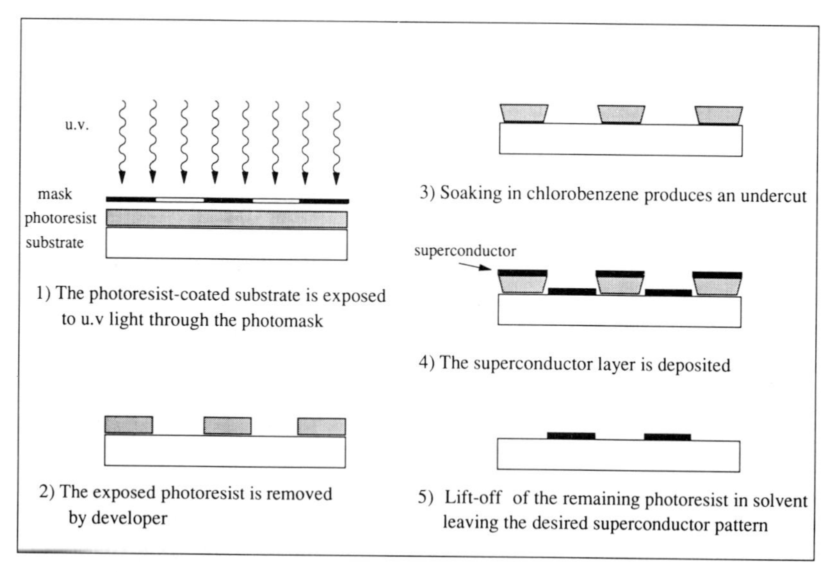

Figure 6.2 Simplified 'lift-off' process for thin film patterning

6.2.3.2 Higher-resolution techniques

Optical lithography is ultimately limited by the wavelength of the radiation used for the exposure process. Diffraction limits the smallest resolvable feature with UV light to around 0.2 μm. In order to make reproducible devices the smallest dimension should be at least a factor of five greater than the diffraction limit, setting 1 μm as the minimum feature size which may be reliably reproduced optically. For even smaller devices, shorter wavelength radiation must be used. Little work has been reported employing x-rays but direct writing electron beam lithography has been used quite widely. The wavelength of an electron with an energy of only 1 eV is as small as 1 nm. In practice the resolution achieved with this method is limited by the minimum electron beam size and by blurring produced by electron back-scatter from the substrate. At present this sets a practical limit of around 30 nm. Without special techniques, back-scattering limits the spatial resolution to the order of thickness of the photoresist ($\sim$100 nm) which is usually polymethylmethacrylate (PMMA). In one of the most straightforward realisations of this process the coated substrate is placed in a scanning electron microscope and the electron beam is passed back and forth under computer control as it exposes complex patterns in the resist without any intermediate stages.

6.2.4 *Tunnel barrier preparation*

The insulating barrier of a Josephson junction must be only about 1 nm thick in order to produce the high critical supercurrent density necessary to minimise capacitative hysteresis (see section 2.2.2). As stressed above, the junction shunt capacitance must be minimised in all high-frequency applications. The exponential variation of critical current density with thickness requires that the tunnel barrier, only a few atoms thick, should be extremely uniform and, most important, continuous. The importance of a homogeneous barrier has become apparent since it has been shown that impurity 'traps' produce localised changes in barrier height (and hence in j_c), causing high leakage current below the energy gap as well as low-frequency noise (Rogers and Buhrman 1984).

For a tunnel junction the tunnelling Hamiltonian formalism, coupled with the BCS theory of superconductivity, predicts that the product of the normal-state resistance R and the critical current i_1 is given by

$$i_1 R < 1.76\Delta/e. \tag{6.1}$$

This derivation ignores the frequency dependence of the supercurrent. However evaluation of the parameter i_1R and the closeness with which equation (6.1) is approached provides a simple measure of junction quality. An additional pragmatic parameter, V_m, has gained acceptance as a figure of merit for junctions (defined as the product of the critical current and the leakage resistance at a bias voltage of 2 mV). V_m gives a qualitative indication of the density of impurity states in the barrier, and values as high as 80 mV have been reported. Figure 6.3 shows a schematic of a typical tunnel junction I–V characteristic.

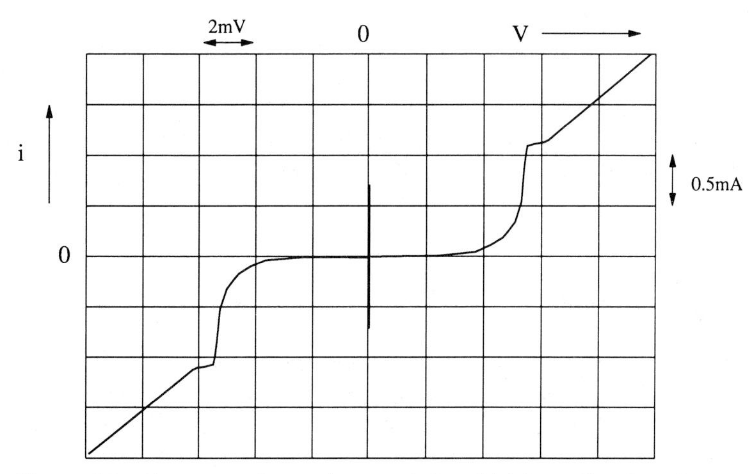

Figure 6.3 Current–voltage characteristic of high-quality Josephson tunnel junction (NbN–MgO–NbN)

Until recently, almost all tunnel barriers were grown *in situ*, in the form of a natural oxide of the base electrode material. This can be brought about by introducing oxygen to the vacuum chamber in which deposition was made, and holding the substrate at an elevated temperature for long enough for the required thickness of barrier to form. Deposition times vary from seconds to hours. A better technique exposes the base electrode to a low-energy radio-frequency-excited oxygen plasma, whose energy and pressure may be controlled to produce self-limiting barrier thicknesses, varying over a wide range. This may use the same sputtering system as is employed for electrode evaporation. A variant of this process replaces the plasma by an ion beam. Another relatively recent development involves the direct deposition of oxide layers (of alumina or magnesia) on highly refractory electrodes such as NbN which, unlike Pb or pure Nb, are tough enough to withstand the high-energy impact of the deposited molecules.

Several theoretical advantages of semiconductor barriers have been apparent for many years. The associated reduction in barrier height means that an equally high critical current density can be achieved with a considerably thicker barrier than for an insulator. For example for Si the barrier height is ~ 0.7 eV compared with 10 eV for some oxides. The tunnel current depends on the product of barrier height and its thickness so that for Si the latter might be as large as 30 nm. This significant thickness might be assumed to make barrier deposition rather simple, as well as reducing the barrier capacitance greatly. In fact it has usually proved necessary to oxidise the semiconductor after deposition since otherwise the junction I–V characteristics show the symptoms of metallic shorts through the tunnel region. Such pin holes can be rendered inactive as effective conduction channels by oxidation.

Si, Ge and InSb barriers have been prepared in this way and a typical reduction of capacitance by about five times has been achieved, compared with Nb_2O_5 barriers. The deposited semiconductors have not yet realised their promise. The necessity of an oxidation process leads one to suspect that deposited layers are intrinsically less homogeneous than oxide layers grown *in situ*.

6.2.5 Outline of an advanced circuit fabrication process

Any attempt to describe all of the multitude of methods developed for tunnel junction fabrication would heavily distort the overall, broad-ranging, approach which this book is intended to take. The preceding sections have indicated how various parts of the fabrication procedure are carried out. In this section a brief summary of just one integrated process is outlined. In this way a flavour of the complex techniques is given without getting bogged down in a welter of detail, which in any case may be found if required in the technical journals.

The process chosen is one developed by Fujitsu in Japan which, at the time of writing, is one of the most reliable processes to have been described. The basic junctions are made from a trilayer of $Nb/Al_2O_3/Nb$ deposited on a whole wafer of Si. Substrates are etched by sputtering in an argon atmosphere before a Nb ground electrode is sputter-deposited. Then an Al layer is put down and an atmosphere of Ar + 10% O_2 is admitted to the chamber which causes a uniform 2 nm thick alumina barrier layer to form. The upper Nb layer is then deposited onto the barrier which is strong enough to remain intact. The resulting trilayer must now be patterned into the device layout required. A photoresist layer is spun on and the ground plane layout defined photolithographically, the minimum feature size being 2.5 μm. Wet anodisation of the ground plane follows. SiO is deposited to provide isolation, holes being subsequently opened through this layer by reactive ion etching (RIE)—see below—through which evaporated Mo resistors make contact. A further layer of SiO protects these. Next a sputtered Nb layer is grown, to provide wiring contacts to the films below. The complete process is summarised schematically in figure 6.4.

Reactive ion etching (RIE) is an invaluable procedure which allows sharp edge profiles to be produced at each stage of the fabrication. A reactive gas, such as CF_4, is introduced into the discharge region which is excited with up to 100 W of RF power. The various materials on the exposed surface of the circuit are etched by the gas at different rates which are both pressure- and power-dependent (the variation in etch rates between different materials may be as great as a factor of ten). In addition, and most importantly, the etch rate for any given material is very anisotropic so that the film is removed considerably more slowly in the direction parallel to the substrate compared to the rate perpendicular to it. This provides steep and reproducible film edge profiles. There is some danger that these sharp edges will cause incomplete coverage by, and lack of continuity in, subsequently deposited layers. Edge cracks may cause unwanted short circuits between non-adjacent layers. All of these problems may be solved by an important new addition to the procedure, a 'planarisation' step. RIE is used to reduce the uneven layers to a flat plane and then a glassy material is spun on, filling any edge cracks, whereupon a further RIE step etches away the glass back to the buried planarised layer, leaving glass only in the cracks. The glass thus smooths off the sharp edges and fills fissures.

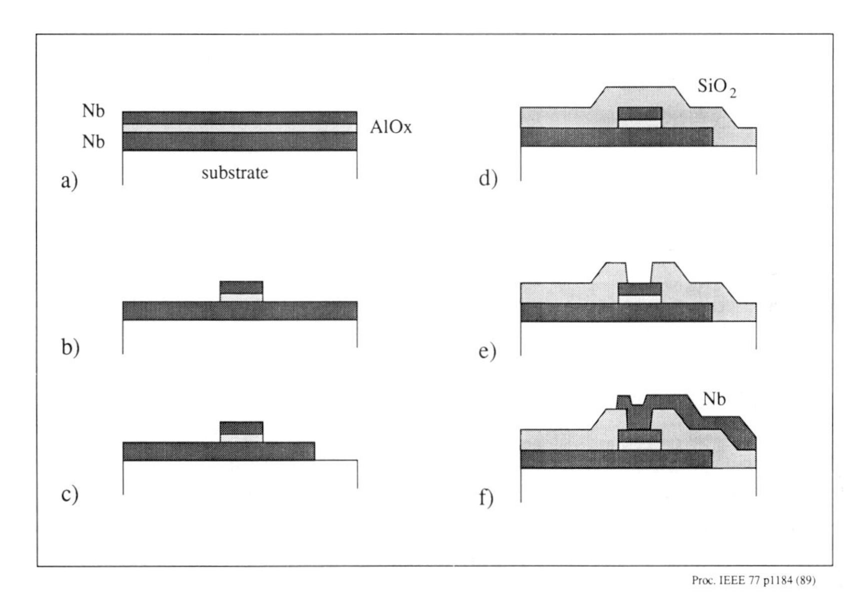

Figure 6.4 Nb/AlO_x/Nb Josephson junction process

The processing outlined here has been used to produce a test large-scale integrated circuit, for example a complete 4 bit Josephson microprocessor clocked at 770 MHz (Hasuo and Imamura 1989). This represents a ten-fold increase in speed over the same design implemented in GaAs technology. Power dissipation is some 500 times lower. Also a 4×4 bit parallel multiplier which uses some 2800 separate Josephson junctions has been described (Kosaka *et al* 1989). This achieved a multiply time of 1.0 ns with a total average power dissipation of only 140 μW. A photomicrograph of a part of this circuit is shown in figure 6.5.

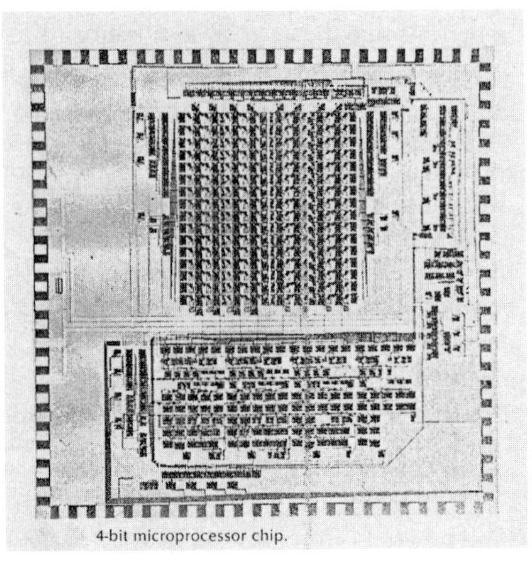

Figure 6.5 View of a 4 bit Josephson microprocessor (Hasuo *et al* 1989)

6.3 High-frequency Detection with Josephson Junctions

We have seen in Chapter 2 that Josephson junctions exhibit extremely non-linear behaviour. Very early in the development of this subject the suggestion was made that this non-linearity could form the basis of a detection method for electromagnetic radiation. Three distinct detector modes have been described which we will deal with in turn in the following sections.

The effect of microwave radiation on the current–voltage characteristic of a Josephson junction has been described by a voltage source model in Chapter 2. The detection of incipient microwave-induced steps in the characteristic due to incident radiation at a single frequency is clearly one way in which such a device might function. Similarly radiation of any frequency incident on the junction will reduce the amplitude of the zero-voltage current step, at least in the low-power limit. These two properties form respectively the basis of the narrow- and broad-band detection modes. In the voltage source model the nth constant-voltage step amplitude varies according to the nth-order Bessel function $J_n(V_1/\Phi_0\omega_1)$ where V_1 is the voltage amplitude across the junction due to the incident radiation at a frequency ω_1. The functional form of $J_0(x) \propto 1 - x^2$ and $J_1(x) \propto x$ for small arguments x (see figure 6.6) implies square-law detection for the broad-band mode and linear detection for the narrow band in the low-signal limits. Although the voltage source model is not generally valid, the more realistic current source model predicts qualitatively the same behaviour. A further complication arises since, in addition to the width of an induced step varying with power, its centre point also shifts relative to the current axis (see figure 2.8).

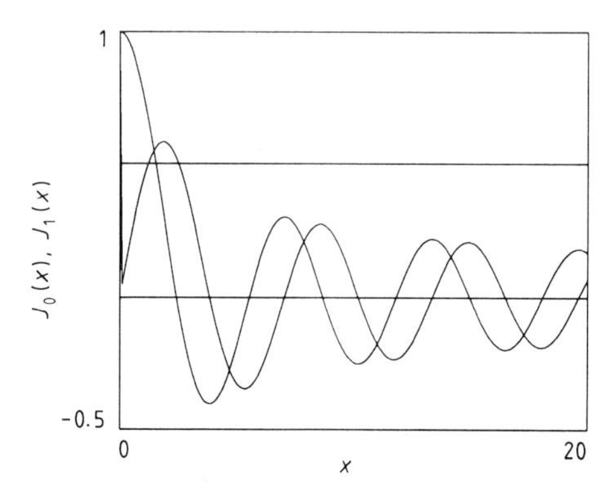

Figure 6.6 Plot of the zeroth- and first-order Bessel functions

6.3.1 Broad-band video detection

A typical experimental detector employs a point-contact junction biased by a constant current into the finite voltage regime, the Josephson frequency corresponding to the bias voltage being much less than the signal frequency ω_1. Incident radiation produces a change in the junction voltage. By chopping the radiation and coherently detecting this voltage change V_1 at the chopping frequency the effect of instrumental drift is minimised. The voltage 'responsivity' S provides a suitable figure of merit for the system

$$S = V_1/P = 2V_1/I_1{}^2R \qquad (6.2)$$

where I_1 is the signal current flowing and R is the junction resistance in the RSJ model so that P is the signal power available at the junction. The voltage swing at the signal frequency can be written

$$V_1 = -(\partial i_1/\partial I_{1\,0})I_1R_d/2 \qquad (6.3)$$

where R_d is the dynamic resistance of the junction at the bias point. For small values of the argument, J_0 is approximated by

$$J_0 = [1 - (V_1/\Phi_0\omega)^2]$$

so according to this idealised model

$$S = R/(2i_1R\Omega^2) \qquad (6.4)$$

Ω being the normalised signal frequency

$$\Omega = \hbar\omega/2eRi_1.$$

The applicability of the voltage source model requires that $\Omega \gg 1$. In practice a point-contact detector provides the simplest way to achieve both a high value of the parameter i_1R and a low value of shunt capacitance C. Theoretical treatments for both point-contact and tunnelling Josephson junctions suggest that the maximum value of the parameter i_1R is related to the energy gap parameter by the expression

$$i_1R \leqslant \alpha\Delta$$

where α is a numerical constant of order 2. The capacitance of a cone of half-angle θ contacting on a conducting plane is well known to be

$$C \simeq \frac{2\pi\varepsilon_0 r}{\ln(\cot\,\theta)} \tag{6.5}$$

where r is the radius of the base of the cone. Microscopic examination of point-contact junctions, particularly after contact has been made, shows that such a model of a well tapered cone making contact with a plane is very much an idealisation. However the absence of hysteretic behaviour in carefully made junctions of this type, with critical currents as low as 1 μA (Stevens 1984), suggests that the capacitance is only of order 1 fF. Inserting this in the above expression yields an estimate of the tip radius of only 0.05 μm. Even the best lithographic methods cannot yet produce features on such a scale with superconducting materials and, furthermore, oxides of niobium in particular have dielectric constants in the range 5–10, thus increasing the capacitance of tunnel barriers over the free space situation. It still appears to be the case that the smallest-area tunnel junctions have a shunt capacitance of order 10 times greater than the best point contacts.

6.3.2 Narrow-band detection

An interesting example of the narrow-band detection technique has been described by Richards and Sterling (1969). A point-contact junction is coupled to a resonant cavity, which produces self-resonant steps in the I–V characteristic. These steps do not have a very low dynamic resistance since the combination of junction and cavity typically shows a Q in the range 10–100. If the junction is biased at a point near one of these steps the spectral response to incident radiation peaks strongly in a narrow band around the resonant frequency of the combination. The fractional frequency width of this response is very much less than $1/Q$, by a factor of at least 30.

6.3.3 Usefulness of Josephson video detectors

Although there are a number of reports in the literature of designs for actual usable video detectors based on Josephson devices, it does not appear that these have been much used in routine applications, attached for instance to radio telescopes. The main reason seems to be that another helium-cooled superconducting detector, the super-Schottky diode, has been shown in practice to be rather more sensitive than any of the Josephson devices, although the theory describing the latter does not support this. Nevertheless a noise equivalent power of 5×10^{-15} W $Hz^{-1/2}$ has been demonstrated for a signal at 90 GHz (Kanter and Vernon 1972) for a Josephson device, a very impressive figure.

6.3.4 Heterodyne detection and high harmonic mixing

The non-linearity of the Josephson effects is also employed in heterodyne detection and mixing modes. Clearly if two external sources of radiation are applied to a junction both sum and difference frequencies will be generated. The particular advantage of Josephson devices over conventional semiconductor diodes is that the sinusoidal current–phase relationship produces a very efficient mixing process and very high-order harmonics of both incident signals can be generated. Thus it has been possible to mix a far-infrared line at 891 GHz in an HCN laser with the 825th harmonic of a microwave oscillator at 1.08 GHz (Blaney and Knight 1974). With conventional mixers it is hard to generate even the 10th harmonic.

Due to the low noise of Josephson devices this mixing process also forms the basis of useful heterodyne detectors which have the particular advantage that the local oscillator power required is very small, and thus especially useful in the millimetre-wave region. Although these mixing and detection techniques are similar in principle to conventional diode applications there is another unique Josephson detection mode which has been developed. This makes use of the alternating supercurrent which flows when a junction is biased into the finite voltage regime. This Josephson current acts as a local oscillator whose frequency can be tuned from zero to the millimetre-wave region by merely adjusting the DC bias voltage.

6.3.5 Mixing and detection with an external local oscillator

Grimes and Shapiro (1968) were the first to demonstrate mixing with Josephson junctions. Since that time a considerable effort has been devoted to trying to understand the process theoretically. In principle all that is required is to solve the following second-order differential equation:

$$(\hbar C/2e)\, \mathrm{d}^2\theta/\mathrm{d}t^2 + (\hbar/2eR)\, \mathrm{d}\theta/\mathrm{d}t + i_1 \sin\theta = I_0 + I_1 \sin\omega_1 t + I_2 \sin\omega_2 t \tag{6.6}$$

where the three terms on the right-hand side of the equation represent respectively the DC bias current and the two applied signal currents at frequencies ω_1 and ω_2. Figure 6.7 shows the I–V characteristic of an Nb point-contact junction for two different local oscillator powers at 35 GHz. The lower diagram shows the mixer output, with maxima coincident with peaks of R_d. The non-linearity of this relationship, so important to the functioning of Josephson devices, has the disadvantage here that it prevents analytical solutions from being found—except for rather trivial cases. As a result theories have concentrated on solutions derived either by numerical integration or analogue modelling. A further complication is that in real high-quality junctions i_1 is frequency-dependent, diverging towards the 'Riedel peak' at a frequency $\omega = 2\Delta/\hbar$ where 2Δ is the superconducting energy gap at the operating temperature (see section 2.5.5). Above this frequency the Josephson current is attenuated approximately as ω^2. A useful figure of merit is the dimensionless parameter α defined by

$$\alpha = (\partial i_1/\partial I_1)\,(R/R_1) \tag{6.7}$$

where R_1 is the assumed source impedances for both alternating current sources. This parameter gives a measure of the change in output current at the intermediate frequency for a given change in the signal current. α is a function of the reduced frequency $\Omega = \omega_1\Phi_0/2\pi i_1 R$ having the values

$$\alpha^2 \sim 1 \qquad \text{for } \Omega \ll 1$$
$$\alpha^2 \sim 0.1/\Omega^2 \qquad \text{for } \Omega \gg 1.$$

The conversion efficiency η characterises mixer performance:

$$\eta = \frac{\text{available power at i.f.}}{\text{available signal power}}$$

where R_d is the dynamic resistance of the junction which in practice is strongly dependent on the bias voltage, thermal noise and incident, unwanted signals. However in practice it is found that

$$R < R_d < 5R.$$

These three equations suggest that, at least for low normalised frequencies, conversion gain ($\eta > 1$) is possible. This has been demonstrated experimentally by Claassen (1975) using a niobium point-contact junction and a signal frequency of 33 GHz. To achieve good η-values at large values of Ω, very high-quality junctions are required and it is vitally important to shield extraneous external signals. Blaney (1978) has suggested that ideal Josephson receiver noise temperatures at 4 K range from 40 K at low frequencies to 430 K at three times the gap frequency. These figures are good compared with the best non-superconducting cryogenic devices. One of the significant advantages of Josephson heterodyne detectors is that they require very low local oscillator powers ($\sim 1\ \mu$W), a particular advantage in the millimetre-wave region where high-power sources are expensive or unavailable.

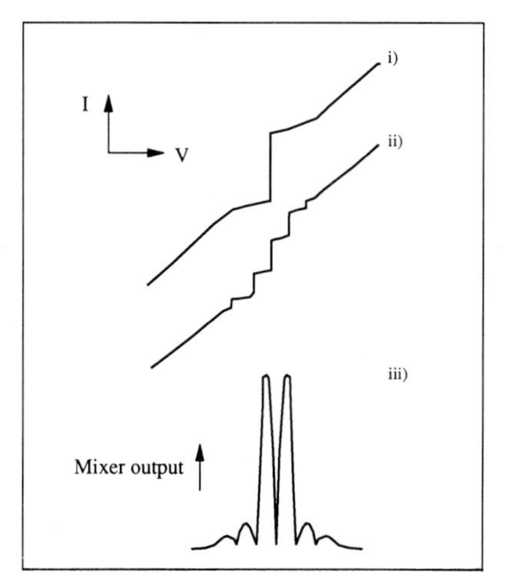

Figure 6.7 *I–V* curve for a point-contact mixer: (i) without microwave power applied; (ii) with sufficient power to reduce critical current to $i/2$; and (iii) mixer output as a function of voltage bias, showing peaks corresponding to the peaks in dynamic resistance

6.3.6 Self-pumped heterodyne detection mode

The feasibility of the detection mode using the Josephson alternating supercurrent as the local oscillator has been demonstrated by Zimmerman (1967). This arrangement has proved useful already in a number of specialised devices such as the frequency modulation noise thermometer (see section 5.9). The advantage of having a local oscillator which can be tuned over a very wide frequency range just by changing the bias voltage is to a large extent offset by the natural linewidth of the oscillator which is typically of the order of tens of megahertz for a junction operated at 4 K. This figure may be reduced by coupling the junction to a resonant structure but only at the expense of losing the broad-band tuning capability. The widening availability of dilution refrigerators providing temperatures of a few mK on a continuous basis at relatively low cost suggests that it should be possible to reduce this linewidth by a factor of around 10^2, to 1 kHz, a much more competitive value, while still retaining the broad-band advantage.

As was mentioned in Chapter 3, the possibility of using a Josephson oscillator to provide the low level of power required for heterodyne detection with Josephson junctions aroused considerable early interest. The maximum power radiated by a single junction into a matched transmission line is of order $4\Delta^2/e^2R$ where R is the shunt resistance of the junction and the linewidth is

$$\delta f \sim 4\pi R_d^{\,2} kT/R\Phi_0^{\,2}$$

where R_d is the dynamic resistance of the junction at the bias point. Junctions with $R = 2\Omega$ have been shown to produce at least 10 nW at 500 GHz. A second approach uses a long Josephson junction to realise a 'fluxon-shuttle' oscillator (Nagatsuma *et al* 1983). Here the junction is penetrated by a linear array of flux quanta when it is cooled in a static applied magnetic field. The Lorentz force between the bias current and each flux line causes the array to move, producing oscillating electromagnetic fields at millimetre wave frequencies. Up to 1 μW of power in the frequency range 100–400 GHz has been coupled into nearby Josephson detector junctions with a linewidth estimated as only 160 kHz, thanks to the low fluxon-shuttle junction shunt resistance.

6.3.7 SIS quasi-particle detectors and mixers

The quasi-particle current, as well as the Josephson supercurrent, in a superconducting weak link shows strong non-linearity (see section 2.2). Low-noise mixers and detectors based on this normal current flow in Josephson junctions have recently come to the fore as the best devices available in the microwave and millimetre frequency range. As with other high-frequency detectors these superconductor–insulator–superconductor (SIS) devices must have small area and capacitance. In fact the present SIS detectors are physically identical to Josephson tunnel junctions. The non-linearity of the quasi-particle current is greatest near the energy gap, corresponding to a voltage of $2\Delta/e$. (Note that the effective charge of the quasi-particles is e, not $2e$.) If a junction is biased near the band gap and a small high-frequency signal is applied, the non-linearity of the characteristic ensures that a change in DC current results which depends on the amplitude of the signal. (An ideal mixer would be switched from zero resistance to an open circuit on each successive half-cycle of the signal.) Figure 6.8 shows schematically the bias points for various superconducting tunnel junction detection modes. This is a broad-band detection process. It is usual to define the current responsivity S in the following way:

$$S = (\mathrm{d}^2 V/\mathrm{d}I^2)/(\mathrm{d}i/\mathrm{d}V) \qquad (6.8)$$

(compare and contrast the voltage responsivity defined by equation (6.3)).

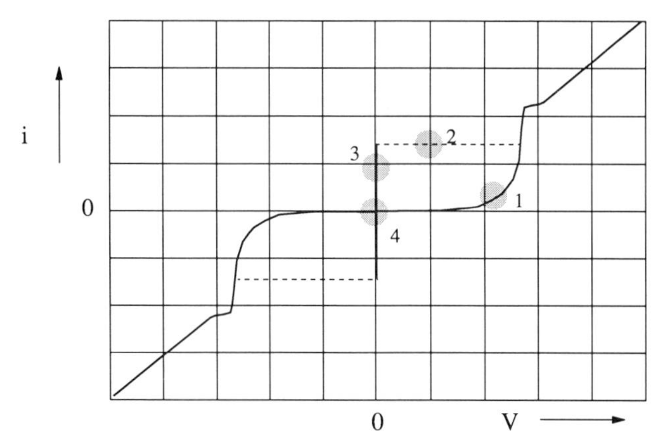

Figure 6.8 *I–V* characteristic for tunnel junction showing hatched regions as bias points for (1) SIS mixer; (2) Josephson mixer; (3) three photon paramp; and (4) four photon paramp (after Chiao 1979)

In the limit for which the resistively shunted junction model applies, it is easy to show that for a mixer without any negative resistance region the maximum value of the conversion efficiency $n = 1$. However it has proved possible to produce devices which are small enough that the curvature of the characteristic is sharp so that $S \sim e\hbar\omega$ where ω is the angular frequency of the applied signal. Then the classical model is no longer applicable and Tucker and Millea (1979) showed that a quantum analysis predicts that conversion gain is possible in SIS mixers.

In practice the limiting value of the usable responsivity is $S < e\hbar\omega$. Such very high values of S, coupled with a low-noise, low-temperature environment and the possibility of conversion gain, give the SIS mixer a significant advantage over any other microwave detector in overall system noise performance at the present time. Noise temperatures as low as 5 K at 35 GHz and 150 K at 300 GHz have been reported.

The major disadvantage of the single-junction SIS mixer is that the device is easily saturated with very low incident power levels ($\sim$pW). The use of linear arrays of n junctions can alleviate this problem by increasing the saturation level n-fold without any significant noise increase. Receivers based on arrays with $n \sim 100$ are now in use at a number of radio telescopes, the first superconducting detectors to enter widespread use in the high-frequency region. These devices outperform not only super-Schottky diodes but also Josephson detectors. This latter fact is not well understood. The Josephson device requires considerably smaller capacitance than would an SIS, complicating the fabrication of thin film Josephson mixers. Also the non linearity associated with the parametric properties of the Josephson supercurrent has been shown in numerical calculations to give rise to increased noise in the region of bias where the detection efficiency is greatest. This effect does not appear with the SIS devices. For this reason the Josephson supercurrent, which is also present in SIS weak links, is suppressed by means of an applied magnetic field.

6.4 Josephson Digital Applications

Superconductivity has had a long association with digital circuitry and computing, as yet without ever having fulfilled its apparent promise in these areas. The earliest serious involvement began in the early 1960s when a large effort was mounted, in a number of different countries, to develop the superconducting cryotron as a logic gate and memory unit. The superficial attraction has been throughout that superconductors can be induced, by applied control signals of one form or another, to switch from the superconducting to the resistive state. This process is strongly suggestive of binary logic. At its most sophisticated the cryotron consisted of a thin film of superconductor overlaid by a control line which, when energised by a direct current, applied a magnetic field to the lower line, great enough to drive it into the normal state. By minimising the size of the thin films the switching speed, as measured by the time to go from dissipationless to resistive states, was reduced to the order of microseconds. A major attraction was that the cryotron produced very little power in the form of Joule heating, less than 1 mW/gate, even when in the 'on' state. In spite of early promise the technology was abandoned without ever being used. Semiconductor device sizes and switching speeds caught up and overtook the cryotron and, coupled with a natural resistance which potential users showed to the idea of liquid-helium-cooled digital circuits, the development programme was stopped around 1965.

Even at that time the beginning of the next honeymoon between superconductivity and the computer industry was already apparent. In 1967 J Matisoo, working for IBM, published a proposal to use the recently discovered Josephson effects as the basis for logic gate operation. Other proposals soon followed, mainly variants on Matisoo's basic idea. He suggested that a direct-current-biased Josephson tunnel junction could be held just below its critical current (with $\langle V \rangle = 0$) until a control current, passed along an overlaid line, applies a magnetic field large enough to reduce i_1 below the bias current, causing switching across to the gap voltage $\langle V \rangle = 2\Delta/e$ (see figure 6.9). This process is strongly reminiscent of the cryotron but has the advantage of much lower power dissipation and essentially higher impedance, so simplifying the matching to conventional electronic components. Since Josephson junctions have much smaller critical currents than do strongly superconducting strips, correspondingly lower control currents are also required.

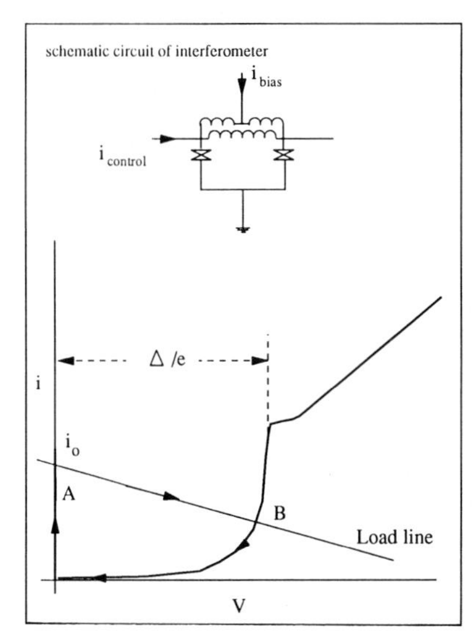

Figure 6.9 Josephson $I-V$ characteristic showing switching from point A to point B when a control current suppresses the critical current

The earliest experiments showed that the switching time of a tunnel junction was not more than 1 ns, four orders of magnitude better than the fastest cryotron of the time. A growing effort was made to develop the new superconducting technology over the next 15 years. The field was dominated by, although not the exclusive preserve of, a large group at IBM laboratories in New York and Zurich. The major step forward came with the realisation that it was advantageous to use two or more very small-area junctions incorporated in a superconducting loop, in place of Matisoo's single junction. This 'interferometer' construction is similar in basic geometry to a DC SQUID (Chapter 4). A loop has the advantage that for a given area the sensitivity to control currents can be made high (high loop inductance) whereas the switching time can be minimised by making the total junction capacitance as small as possible. The two states of such a loop correspond to the bias supercurrent flowing through one or other of the two junctions. The periodic response of the loop critical current means that the same configuration can also act as a memory unit, in which the difference between the two states corresponds to a change of one quantum in the loop flux. Figure 6.10 shows the periodic threshold curve for an interferometer logic gate. In the next section we will deal with the basic dynamics of the switching process between zero and finite voltage states of a single junction, and the current-steering process which determines the time taken for a loop to switch. Subsequent sections describe briefly some examples of gate circuits and system developments.

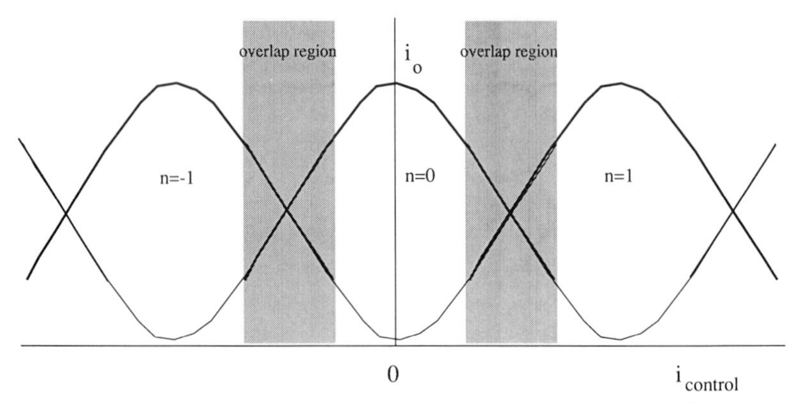

Figure 6.10 Threshold curve of critical current versus control current. Heavy lines represent stable regions. Each lobe is characterised by the fluxon number n

6.4.1 Switching time for a Josephson cryotron

The most obvious determinant of the overall switching speed of the gate is the delay between application of the control current and the appearance of the on-state voltage across the junction. For a junction small in size compared with the Josephson penetration depth $\lambda_J(T)$ the phase difference and current distribution will be uniform across it and consequently a lumped circuit model is appropriate. The simplest approximation is represented by the resistive and capacitance shunted junction (see figure 2.1). The usual second-order non-linear differential equation describes the time evolution of the order parameter phase difference across the junction

$$\frac{C\Phi_0}{2\pi}\frac{\mathrm{d}^2\theta}{\mathrm{d}t^2}+\frac{\Phi_0}{2\pi R}\frac{\mathrm{d}\theta}{\mathrm{d}t}+i_1\sin\theta=i_0. \tag{6.9}$$

Numerical solutions (Stewart 1968) show that the voltage closely follows the behaviour which would be shown by a simple capacitor being charged up by the bias current i_0, provided that the critical current is large enough to satisfy the condition

$$i_1 \gg 2\Delta/Re.$$

Then the approximate time to switch to the gap voltage becomes

$$t \sim 2\Delta C/ei_1. \tag{6.10}$$

Experiments have shown good agreement with this very simple theory. This elucidates why the junction size must be minimised and the critical current density maximised, since the two effects combine to minimise t.

6.4.2 Current steering time

Equation (6.9) can be modified so that solutions provide estimates of the time to steer current into an added load, consisting of a resistive component R_1 and an inductive part L. R_1 should be very small in order to minimise power dissipation occurring in the 'on' condition, in which case the steering time T has a very simple form

$$T \sim \delta i e L / 2\Delta \tag{6.11}$$

where δi is the amount of current steered from junction to load.

6.4.3 Basic memory cell configurations

The memory function based on the storage of a flux quantum in a superconducting ring may be implemented in a wide variety of ways. The simplest possible type involves a single Josephson junction shunted by a superconducting loop (see figure 6.11). The two states of the ring correspond to: (i) no stored flux quantum and a current from an external bias source which flows symmetrically through each arm of the ring ('0' state), and (ii) one quantum in the ring with a net circulating supercurrent ('1' state). To write a '1' state, current is applied to two lines simultaneously which are magnetically coupled to the junction. The effect of the sum of these is sufficient to suppress the critical current below the bias current level, causing the junction momentarily to enter the $\langle V \rangle \neq 0$ state and the current through it switches to the other branch. Removal of all external currents leaves the ring with a circulating supercurrent. A '0' can be written by applying currents to the two control lines without any bias current flowing through the junction. A transition into the $\langle V \rangle = 0$ state now just causes any circulating supercurrent to dissipate, along with the associated stored quantum of flux. Reading the memory state can be done non-destructively, apparently a great simplification in that refresh mechanisms are not necessary. A coincident application of a control current and the gate–junction bias current to a sense junction magnetically coupled to it is sufficient to switch the sense junction into the finite voltage state if a circulating supercurrent is flowing ('1'). In contrast with no circulating current ('0') there is no resulting voltage pulse.

The simplified system described above would not be usable. The main problems with Josephson digital devices are that the margins for successful operation are very narrow and there is no current gain. These two disadvantages can be somewhat alleviated by using small interferometers rather than single junctions; interferometers with as many as four separate junctions per unit cell have been designed (Wada 1990).

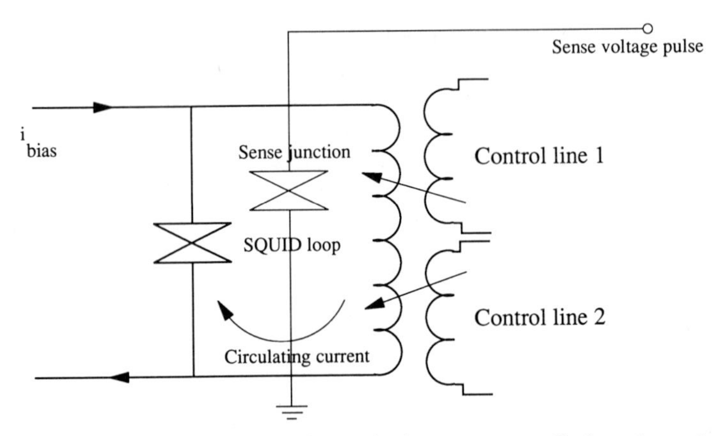

Figure 6.11 A simple Josephson device memory cell, based on the storage of a single flux quantum

6.4.4 Destructive readout

The two-junction interferometer circuit outlined earlier in this section can be used either for logic purposes or for memory storage. The latter involves storing a flux quantum in the ring. In characterising the memory cell a threshold curve of bias current as a function of net control current is useful (see figure 6.10). The critical bias current falls as the control current increases, with an approximately straight-line dependence, reaching a minimum and then rising again to a maximum. The threshold curve is flux-quantum periodic of course, with overlapping curves corresponding to each included additional flux quantum. Writing to this memory cell involves three control lines, one of which maintains the bias point within the overlap region of the threshold curve (see figure 6.12). Application of the two other currents is sufficient to switch the selected junction into one state or the other depending on the sense of the control currents (unselected cells sense only one of these currents, which on its own is not sufficient to switch its state). The interferometer bias current (I_y) is applied before the control currents are removed. Reading the cell involves applying I_y first, followed by the control currents in a particular (positive, say) direction. If the control currents change, the cell switches to the $\langle V \rangle \neq 0$ state and this may be detected. Unfortunately in this process the contents of the memory are lost and must be rewritten.

6.4.5 Magnetically coupled logic gate operation

Essentially the same SQUID interferometer circuit as described above as a memory cell can also form the basis of a logic gate. In this case the important function is not the storage of a flux quantum in the ring but rather the steering of current from one path to the other, the two paths corresponding to the two possible output states of the logic gate. A recurrent problem of Josephson digital circuits is their relatively low available gain. In principle the gain from a single gate could be made as large as desired by biasing the device arbitrarily near to its threshold. Then an infinitesimal control current will produce a finite output current change. The problem remains that it is not possible to match the critical parameters of all the individual gates of the total circuit with arbitrary precision, so that the reproducibility of the fabrication process sets an effective limit on the gain available. Typically a gain of about three has been used as a design figure where a spread in critical current values is around 10%.

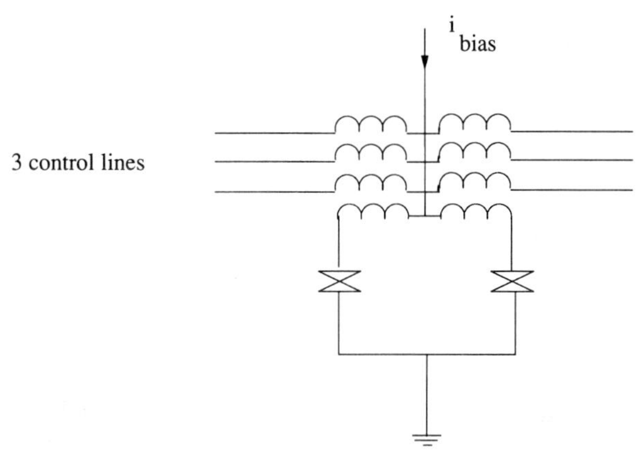

Figure 6.12 Schematic of two-junction memory cell with three control lines

With an interferometer configuration there is the possibility of using multi-turn control line coils to give higher gain. The disadvantage is that the increased inductance of the coil (and incidentally the ring inductance itself) tends to slow down the current steering time (as determined by equation (6.11)). The spatial extent of an interferometer ring also means that the packing density of circuits cannot be as high as for single-junction gates. However the input to output isolation of the SQUID geometry is almost perfect, a significant advantage as we will see in the next section.

6.4.6 Direct coupled gates

It would of course be possible to cause a Josephson junction to switch into the finite voltage regime by direct injection of current into the bias circuit. The problem with this is that the output current flow is hardly isolated at all from the input side and this output current would then influence the input states of other gates in parallel. The attraction of 'direct coupling' can be achieved, however, by including so-called 'diode' junctions. Figure 6.13 shows the bridge arrangement of two junctions and two resistors and the associated threshold curve. With no input signal current the bias current is set so that both junctions are in the zero voltage state. An input causes junction J_2 to switch, the current being diverted into the other arm of the bridge, which in turn switches into the finite voltage state. The high normal-state resistance of this latter 'diode' junction (of order 100 Ω) ensures reasonable input/output isolation. Typical resistance values are $R_1 = 1\ \Omega$, $R_2 = 0.5\ \Omega$ and $R_0 = 12\ \Omega$. The size of this device is typically three times smaller than for an interferometer logic gate.

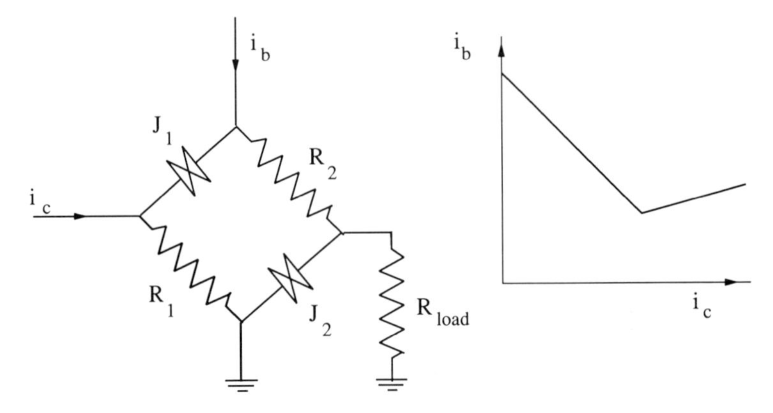

Figure 6.13 Direct coupled logic 'bridge' with threshold curve

6.4.7 Prognosis for the Josephson computer

IBM gave up development of the Josephson computer in 1983 after a very significant investment, on the grounds that the circuit yields were not good enough. The tolerance in circuit parameters achieved with Pb–In alloy technology was not good enough to attain the expected advantages. In addition the switching speeds of single Ga–As devices had been shown to be comparable with that of Josephson gates, although the latter dissipate some 10^4 times less power per gate.

Interest in the Josephson computer is now entirely concentrated in Japan, where a coordinated programme is going on at a number of university, government and commercial institutions. Considerable progress has been made in implementing an all-refractory junction system with higher tolerances than achieved by IBM. A complete 4 bit microprocessor design has now been realised (Kotani *et al* 1989), as have operational 1 kbit RAM chips (Wada 1990). It seems clear now that a very high-speed Josephson computer can and probably will be made. It is, however, still too early to say whether or not the Josephson computer will ever be realised in large numbers.

6.5 Analogue to Digital Conversion and High-speed Sampling

It is to be expected that any technology which is appropriate to high-speed logic gate implementation will also show promise in applications involving the interface between digital and analogue signals.

6.5.1 Analogue to digital conversion

The high bandwidth and fast switching speeds available with Josephson junctions make attractive the possibility of constructing AD converters based on superconducting technology. An added advantage is that the periodic nature of the SQUID response or the exact linearity of the Josephson voltage–frequency relationship provides an accurate scale which provides a reference for the conversion. A proposal to make a very accurate converter based on equation (2.25) involves applying the analogue signal across a Josephson junction as a voltage source. Then the supercurrent oscillation frequency f (which may be as high as 100 GHz) is measured with high accuracy. At low temperatures and with care it is possible to measure this frequency to 1 in 10^6, giving up to 20 bit accuracy. The measurement is not particularly fast since the conversion is essentially serial in nature; to get extra bits of resolution it is necessary to count longer. However the very high bit number available may encourage development of this device for its unique properties.

6.5.2 SQUID AD converter

An entirely different arrangement has been developed by Hamilton *et al* (1985). The device is centred on two identical thin film SQUID interferometers of small inductance (20 pH). The signal is magnetically coupled to each of them. In addition they are biased in series by a current fed from a clocking circuit. This produces a very sharply rising current edge when two Josephson junctions in series are caused to switch into the finite voltage state. If the signal current at this clock instant happens to be at a value corresponding to a ‘1’ region shown on the threshold curve diagram (figure 6.13) the reference SQUID will switch, producing a ‘1’ at the output. Alternatively if the signal current lies in the ‘0’ region the signal SQUID will switch, reducing the bias across the reference SQUID to the level where this interferometer will not switch, giving a '0' at the output. This deals with a single bit conversion. More significant bits are converted by including parallel sets of SQUID pairs coupled to the signal by resistive divider circuits, each reducing the signal by a further factor of two (see figure 6.14). A 6 bit converter has been built and demonstrated with a 10 ns conversion time. The potential performance of the device suggests much higher conversion rates, up to 5 Gsamples/second, since the prototype has already demonstrated 6 bit conversion in 100 ps.

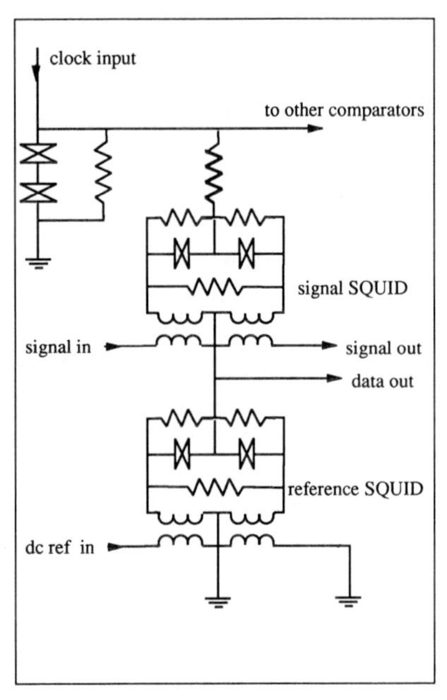

Figure 6.14 Circuit diagram of a SQUID comparator used in an ADC (see Hamilton *et al* 1985)

6.5.3 High-speed Josephson sampler circuit

A sampler is a device which allows a fast but accurately repeated waveform to be sampled with high time domain resolution. The heart of the fastest reported Josephson system is a single small-area tunnel junction to which is directly applied the signal (see figure 6.15). In addition a superconducting pulse generator and variable delay is provided on the same chip. The delay is swept over a range corresponding to the extent of the signal waveform which is desired to be viewed. At each operation of the trigger pulse the sum of the trigger and the signal is applied across the sampling junction. A fast conventional semiconductor comparator circuit senses the voltage across the sampling junction and after high-gain amplification this is fed back to the bias current lead of the sampler, with the result that it is held in the 50% on condition. The feedback current is fed to the *y* plates of an oscilloscope and a voltage proportional to the delay is fed to the *x* plates. As the delay on the trigger signal is steadily increased, the sampler junction samples a later piece of the waveform and gradually, over many cycles, builds up a complete picture. The device described by Wolf *et al* (1985) is the fastest yet reported with a resolution of 2.1 ps and has been used to examine the switching transitions in Josephson logic interferometers. Elaborate electro-optic samplers have demonstrated sub-picosecond resolution but in comparison the Josephson sampler on a single chip is a very simple and compact alternative, besides the fact that further improvements in design should allow significant speeding up of the Josephson device.

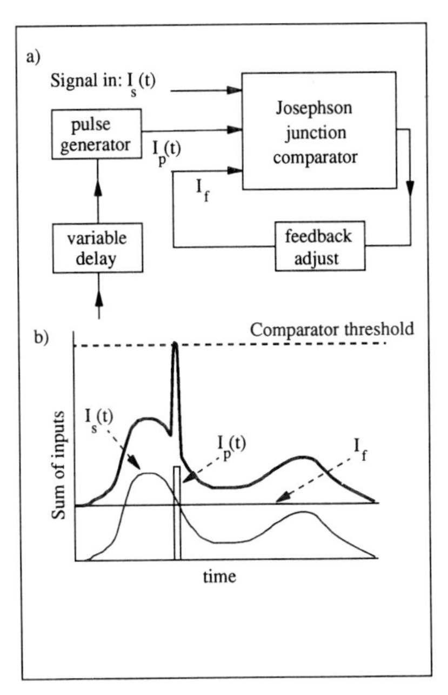

Figure 6.15 (*a*) A Josephson sampler circuit schematic with (*b*) the sum of the variable delay pulse, feedback and signal voltages just exceeding threshold

7

A Practical Guide to Using SQUIDs

Much has been written, in this book and elsewhere, about the extreme sensitivity attainable with devices based on superconducting electronics. Anyone who has attempted to make use of this unequalled sensitivity will know that the gulf between first attempts and final success is often considerable. This generally arises from a lack of appreciation of the essential precautions needed to shield the SQUID from extraneous interference and noise, precautions which far exceed what is necessary with less sensitive conventional electronics. As a realistic counterweight to the somewhat rose-coloured view which earlier chapters may have given, we will try to give some practical hints on how to make real measurements in real environments with SQUID systems.

Possible sources of interference may be simply listed and will be dealt with in what is frequently their order of importance. Magnetic fluctuations may be expected to be dominant, but mechanical noise is also difficult to deal with, and electromagnetic, electrical and thermal instabilities must be carefully considered. It would be an oversimplification to assume that all these sources are independent.

7.1 Magnetic Noise

At most sites on the earth's surface the ambient magnetic field is in the range 30–100 μT, and, even in the absence of man-made electromagnetic interference, it fluctuates (with a rather unpredictable frequency spectrum below about 100 Hz), and has a typical RMS amplitude of ~ 1 nT. These fluctuations arise mainly from changes in the geomagnetic ring current in the radiation belts high above the earth's surface.

Since a SQUID with a multi-turn pick-up coil can detect a change in flux density of only 10 fT, it is essential that the input coil be carefully shielded against fluctuations of the earth's magnetic field. The attenuation factor must be at least five orders of magnitude, even at a magnetically quiet site. In Chapter 5 we discussed how gradiometer coil arrangements discriminate against distant sources of magnetic noise. If a SQUID is to be used to sense a local source at room temperature then such a multi-coil discrimination arrangement will be necessary. Furthermore the system will almost certainly have to be operated at a magnetically quiet site, far from mains-powered devices since these would represent rather close and strong sources of fluctuating field amplitudes at 50 or 60 Hz in a normal laboratory situation.

Of course it may be possible to enclose the system in a magnetically shielded region. A few such very low-field enclosures have been built at great cost, using as many as six alternate layers of mu-metal and copper (Mager 1982), and have been particularly useful in biomagnetic studies using SQUIDs. A cheaper solution which may be as effective is to use a single layer shield of 1 cm thick aluminium sheet (Zimmerman 1977). Such a Faraday cage will provide electromagnetic shielding down to frequencies for which the skin depth becomes comparable with the wall thickness (around 10 Hz for the above example). Other workers have reported (Prance *et al* 1985) that many layers of silicon–iron sheet, as used in power transformer cores, applied to the exterior of both cryostat and room temperature instrumentation provide cost-effective magnetic shielding. Yet another cheap solution is to get a central heating contractor to build an 'oil tank' around the entire SQUID system, with an electrically screened door, sealed around its edges with copper spring 'fingers' to make low-resistance contact. Domestic oil tanks are made from mild steel rather thicker than that used in commercial screened rooms so that the performance of the latter may be equalled if care is taken to provide the door with a good electrical seal (Stevens 1984).

If experimental conditions allow, by far the best magnetic shielding is provided by enclosing the sensing coils and the signal source inside a superconducting shield. For a completely closed shield the Meissner effect ensures that no change in magnetic field outside the box can be communicated to the interior unless the critical field of the superconductor is exceeded. Furthermore even for a shield with some holes, external field changes are rapidly attenuated, with an exponential dependence on the distance from the hole, on a scale set by the hole radius. In principle it is possible to build an environment of any required size, which is at room temperature but shielded by a superconductor. All that is needed is a reentrant cryostat to contain the shield. The cost of such an enclosure of sufficient size to contain say, a person, will be rather high so that at present this solution has not been attempted. The widespread use of whole-body NMR scanners using superconducting magnets shows that this size scale would also be possible for a SQUID system and may become cost effective if biomagnetism is seen to have sufficient diagnostic significance.

7.1.1 *Shielding by means of an open-ended tube*

When a cylindrical open-ended tube is cooled through its transition temperature in an external magnetic field the resulting trapped field through the tube can have any value, in principle, between zero and the value of the external field. If the transition through T_c is well controlled and the tube goes superconducting smoothly from one end to the other then, provided that the tube cross section is also uniform along its length, it is found by experiment that the trapped field is approximately equal to the external field and very spatially uniform. In this section we will examine the functional dependence of this variation $dB(z, r, \theta)/dB_{ext}$ on z, r and θ, the coordinates of an internal point.

Consider a semi-infinite tube of superconductor with radius a and negligible wall thickness, extending parallel to the z axis in the positive z direction. Imagine the magnetic field is initially everywhere zero and that an external field, uniform at large distances from the tube, is then applied. What will be the magnetic flux density within the tube? In Chapter 1 the relationship between magnetic vector potential and supercurrent density was derived (equation (1.18)):

$$\boldsymbol{j} = -(4e^2\alpha/m\beta)\boldsymbol{A}. \tag{7.1}$$

Taking first the region internal to the tube ($r < a$) we see that here $\boldsymbol{j} = 0$ so that $\boldsymbol{A} = 0$ and $\boldsymbol{B} = \nabla \times \boldsymbol{A}$. The latter two relationships combine to show that in this case $\boldsymbol{B}$ can be defined as the gradient of a scalar potential φ since for any scalar

$$\nabla \times (\nabla\varphi) = 0.$$

When $\boldsymbol{j} = 0$, φ satisfies Laplace's equation

$$\nabla^2\varphi = 0.$$

In addition we know that the component of the magnetic field normal to the superconducting tube vanishes everywhere on its surface. For the cylindrical geometry and coordinate choice shown in figure 7.1 the general solution for φ is given by

$$\varphi(r, \theta, z) = \sum_m \sum_n J_m(k_{mn}r)\exp(-k_{mn}z) \times [A_{mn}\sin(m\theta) + B_{mn}\cos(m\theta)] \tag{7.2}$$

where m and n are integers, and J_m is the mth-order Bessel function, k_{mn}

being the nth root of the equation

$$\partial J_m(k_{mn}r)/\partial r = 0.$$

We are interested in how the interior magnetic field reduces in magnitude as we go into the tube from the open end. For a uniform external field parallel to the z axis the geometry of the situation demands that only $m = 0$ solutions, which have complete rotational symmetry about the z axis, are possible. In this case the smallest value of k_{0n} corresponds to the $n = 1$ solution for which $k_{01} = 3.83$. This sets the slowest rate at which the field falls off with z when $z \gg 0$:

$$B_z(z) \propto \exp(-3.83z/a). \tag{7.3}$$

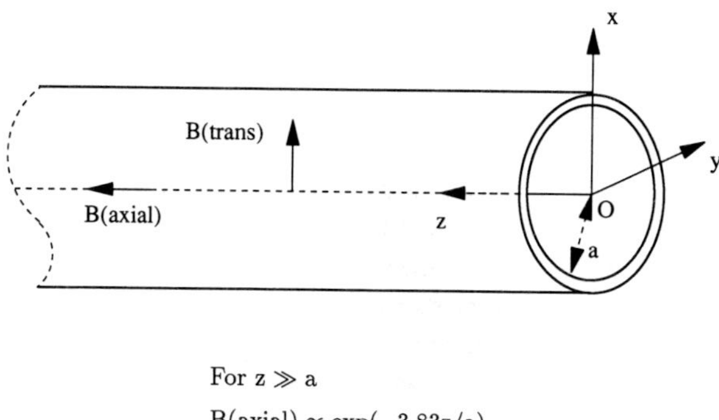

Figure 7.1 Shielding of magnetic fields by semi-infinite superconducting tube

Thus for example the field is attenuated by a factor of 10^4 within a distance of only three radii from the tube end. For transverse fields the slowest dependence on z is as $\exp(-1.84z/a)$ so that in this case the field is attenuated by the same factor within six radii. It is thus the exponential attenuation of external fields which makes the superconducting shield so effective. The attenuation is brought about by a supercurrent flowing around the inner surface of the tube near its end, whose density $j(z)$ diverges as $z \to 0$.

For the case of a long open-ended tube there is also a set of possible solutions in which the interior field approaches a constant value far from the ends, with the added boundary condition that the total magnetic flux through any cross section is constant, being an integer number of flux quanta. The flux density $B(r, \theta, z)$ at any point may be readily found by numerical methods, based on the requirement that $\Phi = n\Phi_0$. Again the persistent current density $j(z)$ diverges towards the tube ends and at some point very close to the end will exceed the critical current density of the superconductor, which condition prevents any further increase. This however is a very small perturbation which can be ignored for all interior points more distant than about one radius from the open end.

7.2 Mechanical Construction of SQUID Circuits

We shall see in the following chapter of this book that SQUIDs have been used to produce the most sensitive accelerometers and displacement transducers which are presently available. The converse of this is clear, namely that these devices must be very carefully screened from unwanted mechanical noise. For conventional SQUIDs there are only two basic mechanisms which can couple them to mechanical movement. The first, and probably the most important, arises from movement of the SQUID or its input coils relative to a non-uniform magnetic field distribution, leading to unwanted output changes. The second is direct changes of coil dimensions, or even of the area of the SQUID ring itself, associated with applied mechanical stress. This could be a quasi-static effect or arise from acoustic waves propagating through the system, perhaps initiated by random mechanical impulses. This is generally a much smaller effect than the first and may be dealt with by the same methods, which we next consider.

There are two lines of attack in minimising the first source of noise. Obviously the structure which positions the SQUID and its associated flux transformer with respect to the shielding should be as rigid as possible. High rigidity is sometimes difficult to achieve as there are a number of cryogenic constraints on materials which may be used (see section 7.3). The requirements of the system may dictate that non-magnetic or even non-metallic materials should be used wherever possible. Particular care must be taken to hold the RF and signal coils fixed with respect to the superconducting ring. In this respect, at least, the toroidal RF SQUID structure is clearly superior to the two-hole type, in which the two coupling coils on non-metallic formers are a sliding fit into the two holes. An entirely planar SQUID with planar coupling coils all mounted on a common substrate might be expected to be the best arrangement. Insufficient published experience with such systems is at present available to determine whether this is so in practice.

A second approach to reducing mechanical noise is to make the trapped field in the neighbourhood of the SQUID and its coil as spatially uniform as possible. Surprisingly little attention is usually paid to this approach, but if used in conjunction with rigid construction methods, the benefits are great. As well as making the trapped field as uniform as possible it is generally desirable to try to minimise the magnitude of the field, thereby reducing the absolute value of the gradient.

External ferromagnetic shields provide a simple way of reducing the ambient magnetic field to a low level. For example a three-layer mu-metal shield can readily produce regions of field some one thousand times less than the earth's field, perhaps of order 10 nT. On its own this is not enough. It is not widely appreciated that the way in which a superconducting shield is cooled through its transition temperature is also very important. Indeed if a shield is cooled rapidly in a non-uniform way in a very low external field, thermoelectric currents resulting from a temperature gradient can generate much larger fields which may become trapped and, when surrounded by superconducting regions, compressed as these regions propagate through the material. Such compression and trapping of flux will clearly give rise to very non-uniform gradients. It is best when designing a shield system in a cryostat to require that it cools slowly from one end. This may be achieved by enclosing it inside a vacuum jacket and providing a well defined thermal link from the liquid helium bath to an appropriate point on the shield. A less complex arrangement is to add some thermal lagging on the outside of the shield (foamed plastic cut to shape or cast *in situ* may be used), a small area being left unlagged so that it will go superconducting first. This area should obviously be at one end of the structure. The homogeneity of the shield's construction is also important in deciding how uniform the trapped field will be. Thus it is preferable to avoid welded or soldered seams as far as possible. Although complex shapes can be readily made by cutting and soldering lead sheet this type of shield is certainly not ideal for applications requiring the highest sensitivity. Structures machined from solid bars of niobium are probably the best. Good results have also been reported using shields cast from molten lead. Little has been published on field uniformities achieved with various types of shield. The author has found that with a 60 mm diameter soldered shield typical gradients of around $10^{-7}\,\mathrm{T\,m^{-1}}$ may be expected. Inside a homogeneous shield the gradients may be three orders of magnitude lower. Thus to keep the mechanically generated noise less than our typical system noise of $10^{-4}\Phi_0\,\mathrm{Hz}^{-1/2}$ (or $\delta B \sim 10^{-14}$ T in field terms) requires that the relative displacement of coil and a soldered shield must be less than 0.1 μm.

7.2.1 Noise and the London moment

SQUIDs can also couple directly to rotational motion. In the discussion of flux quantisation in Chapter 1 an expression (equation (1.13)) was given for the gauge invariant phase of the order parameter θ

$$\nabla\theta = mj/2e\hbar + 2eA/\hbar. \tag{7.4}$$

When deriving the condition for flux quantisation in a strongly coupled ring a contour was selected deep within the material where the current density $j = 0$. The rotation of a ring with angular velocity ω produces an effective circulating supercurrent density $\boldsymbol{j} = 2e\boldsymbol{\omega} \times \boldsymbol{r}$ at a point a distance r from the origin of a non-rotating frame. Since in this situation there is no contour within the ring which has zero current density, we see that there will be an additional contribution to the phase change around the ring arising from the first term on the right-hand side of equation (7.4). This amounts to

$$\delta\theta = (m/\hbar)\int \boldsymbol{\omega} \times \boldsymbol{r} \cdot \mathrm{d}\boldsymbol{l} = (m/\hbar)\int \boldsymbol{\omega} \cdot \mathrm{d}\boldsymbol{S}. \tag{7.5}$$

Adding $\delta\theta$ to the flux quantisation condition shows that as a result of the rotation the flux linking the ring is changed by an amount

$$\delta\Phi = (2m/e)\boldsymbol{\omega} \cdot \boldsymbol{S}. \tag{7.6}$$

This rotationally generated flux, known as the London moment, will exist in the SQUID ring itself as well as in any superconducting circuits coupled to it. Thus mechanical vibrations which give rise to unpredictable rotational motion are a source of noise. For a typical pick-up coil with 100 turns and a cross-sectional area of 10^{-4} m^2 the angular velocity must be kept below 10^{-5} rad s^{-1}.

7.3 Noise Sources Arising from Temperature Fluctuations

As should be clear by now, there is no simple division of noise sources into independent categories since the generation of noise often results from an interaction of several sources. So, for example, the most obvious source of temperature fluctuation noise comes from the change of mechanical dimensions of objects as a result of a change in temperature. Fortunately it is an experimental fact, also predicted by the third law of thermodynamics, that coefficients of thermal expansion tend towards zero as the temperature approaches 0 K. Typically in the liquid helium region the thermal expansion coefficient for metals $\alpha \sim 10^{-8}$ K^{-1}. For plastics the values are perhaps ten times greater whereas for ceramics and glasses α may be typically ten times smaller. In addition temperature control at liquid helium temperatures is simpler than at room temperature so that a system based on maintaining the vapour pressure constant above a bath of liquid helium can easily produce a stability of better than 10 μK. (Recent reports suggest that use of the transition temperature of a pure unstrained superconductor as a reference point allows temperature stability to ± 10 nK and the dissipationless superconducting kinetic inductance bolometer (McDonald 1987) is predicted to improve on this by a further three orders of magnitude.) In spite of these advantages arising from the cryogenic environment, it is still necessary to design the SQUID circuit to minimise thermal expansion effects. In the last section we estimated that the SQUID circuit must not move by more than 0.1 μm with respect to the shield. Assuming a temperature stability of 1 mK and a variation in $\alpha \sim 10^{-8}$ between constructional materials used, the overall size of the device should be less than about 10^4 m, easy enough to satisfy!

Another source of thermal noise arises from the variation of superconducting properties with temperature. The critical currents of all Josephson junctions are temperature-dependent to varying degrees. Most affected are microbridge weak links which exhibit an exponential rise in critical current as the temperature falls:

$$i_c(T) = i_c(0)\exp(-2T/\Delta(0)).$$

Tunnel junctions show the same dependence on temperature as does the superconducting energy gap for $T < T_c/2$ i.e. $\Delta = \Delta(0)\,[1 - \exp(T/T_c)]$ so that variations in i_c are very small. Point contacts do not seem to be well characterised but can show behaviour anywhere between the limits of tunnel junctions or microbridges. Even for tunnel junctions it is necessary to operate at a temperature less than about $T_c/2$ if thermal effects are to be insignificant as a noise source. A variation in the critical current of an RF device would alter the slope of the V_{RF} versus I_{RF} curve (figure 3.7). Similarly for a DC SQUID a variation of the critical current of either junction will change the direct voltage appearing across the parallel combination, thus appearing as a simple source of voltage noise.

Another parameter which is temperature-dependent is the superconducting penetration depth λ which can convert thermal fluctuations into equivalent flux noise. Thus a change in λ will produce a corresponding change in the inductance of the SQUID ring or of any coil coupled to the SQUID itself, or even of any superconducting shield with non-negligible magnetic coupling to the SQUID. This effect may be seen as arising simply from the increased magnetic area of a loop as λ increases, or equivalently as additional kinetic inductance associated with the changed density of pairs near the superconductor surface. Taking the simple London theory as a basis, the penetration depth varies as

$$\lambda(T) = \lambda(0)\,[1 - (T/T_c)^4]^{-1/2} \tag{7.7}$$

so that the rate of change with temperature becomes

$$\mathrm{d}\lambda/\mathrm{d}T = 2\lambda(0)\,[1 - (T/T_c)^4]^{-3/2}(T/T_c)^3. \tag{7.8}$$

This function is plotted in figure 7.2. A simple relationship exists between the penetration depth and the surface reactance per unit area X_s:

$$X_s \approx \mu_0\omega\lambda(T).$$

A simple equivalent circuit description shows that X_s is added in parallel to the normal geometric ring inductance, giving an effective total inductance

$$L_{\text{eff}} = L_1 + k^2 X_s/\omega = L_1 + k^2\mu_0\lambda(T) \tag{7.9}$$

where k is the coupling parameter between SQUID ring and shield. Since the output voltage to input flux transfer function $\partial V/\partial\Phi$ is proportional to L_{eff}^2, any variation in λ will cause a consequent change in output level $\langle V\rangle$ with no change in input signal (in other words noise). We can write for the output variation with T

$$\mathrm{d}V/\mathrm{d}T = (\partial V/\partial L_1)(\mathrm{d}L_1/\mathrm{d}T) + k^2(\partial V/\partial L_s)(\mathrm{d}L_s/\mathrm{d}T).$$

Equation (3.5) in section 3.3.1 shows that

$$\partial V/\partial L_1 = V/2L_1 \qquad \partial V/\partial L_s = V/2L_s$$

and from the dependence of L_1 and L_s on λ we deduce that

$$\mathrm{d}L_1/\mathrm{d}T = [2L_1/(r_1+\lambda)](\mathrm{d}\lambda/\mathrm{d}T) \qquad \mathrm{d}L_s/\mathrm{d}T = [2L_s/(r_s+\lambda)](\mathrm{d}\lambda/\mathrm{d}T)$$

where r_1 is the coil radius and r_s is the shield radius.

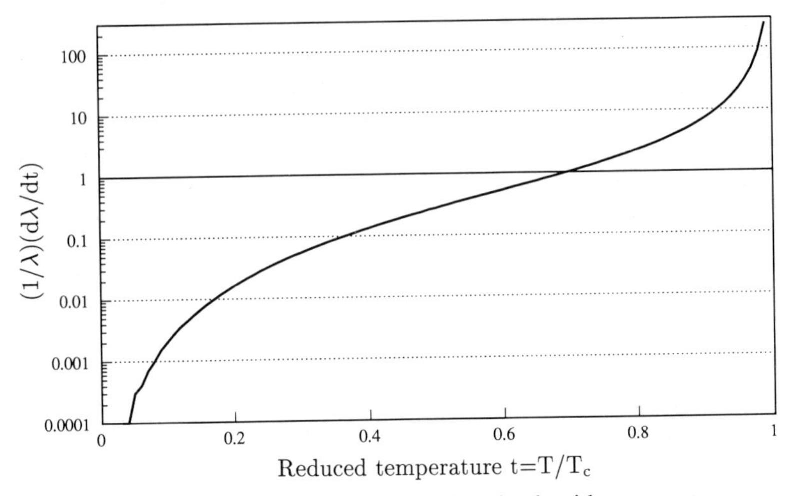

Figure 7.2 Variation of the penetration depth with temperature

Combining these it is apparent that the variation of $\lambda(T)$ is much more important in the signal coil inductance than that of the SQUID ring itself since L_1 is typically 10^2 times greater than L_s whereas $r_s \geqslant r_1$. If L_1 is of the order of a few microhenrys, the surface inductance contribution will be of the order of 0.1% of the total. To maintain the equivalent input flux noise below the intrinsic level of around $10^{-4}\Phi_0\ \mathrm{Hz}^{-1/2}$ requires that

$$[\mu_0\Phi(\mathrm{d}\lambda/\mathrm{d}T)/L]\,\delta T < 10^{-4}\Phi_0$$

where Φ is the total flux through the coil. Thus $\Phi\delta T < 10^4\ \Phi_0$ at a temperature of $T_c/2$, showing that there is a trade-off between the maximum tolerable temperature fluctuation and the flux through the loop. Here is yet one more reason to minimise trapped flux in SQUID ring and flux transformers. It is of course possible in principle to include a small heater in contact with any superconducting coil, allowing it to be driven normal to free any trapped flux generated by thermal inhomogeneities on first cooling.

7.4 Electromagnetic Interference

SQUIDs respond with enormous sensitivity to electromagnetic fields from DC up to millimetre-wave frequencies. Great attention must be paid to screening superconducting electronic devices throughout this frequency range. We have seen that a SQUID input coil can be adequately shielded against low-frequency magnetic field fluctuations by enclosing it in a superconducting screen. Such a shield will clearly also screen out higher-frequency electromagnetic fields, at least up to frequencies for which the wavelength is comparable with the linear dimensions of the largest aperture in it. Unfortunately in practice this assumption turns out to be something of an idealisation. In almost all applications it is necessary to pass wires through the shield and these can act as antennae which couple in external fields. For a typical cryostat these leads are of such a length that coupling will be a maximum for radio-frequency fields. If the leads are only required to introduce low-frequency signals into the space inside the shield it is a relatively simple matter to include appropriate low-pass filters, preferably close to the outer wall of the shield. For audio-frequency signals simple R–C filters formed by ceramic RF feed-through capacitors and 1 kΩ metal film resistors work very effectively, both components operating reliably down to $-273\ ^\circ$C.

For an RF SQUID another source of coupling of interference is via the RF line itself. Ideally one should use the best coaxial cable with a continuous outer screen, many skin depths in thickness. In reality this cannot be used since the thermal load it would introduce would be intolerable in most circumstances, resulting in an excessive rate of evaporation of the cryogenic fluid (liquid helium in general) which is providing cooling. Instead a cable with a loosely woven copper screen is installed inside a rigid tube of stainless steel, of such a wall thickness as to be greater than the skin depth of that material in the radio-frequency range ($\sim$0.03 mm at 20 MHz). All joins in leads and shields must be of very low resistance, being soldered or welded where possible, or otherwise made with tightly screwed connections.

Even when all these precautions are taken it is still advisable to keep any sources of RF interference as far as possible from the SQUID itself. Particular attention should be paid to digital electronic instruments such as microcomputers, many of which radiate strongly in the frequency range from about 1 to 20 MHz. If digital instruments must be connected to the SQUID output for purposes of data collection or control, then if possible opto-isolation should be used in all leads.

7.4.1 Elimination of mains frequency interference

In a typical laboratory environment the electric field strengths at mains frequency may be as high as $10^2\ \mathrm{V\,m^{-1}}$. The spectral analysis of any SQUID system almost invariably displays a prominent peak at 50 Hz and some of its harmonics, even when, as in most commercial systems, a notch filter at this frequency is included. The efforts required to reduce this interference depend very much on what ultimate system sensitivity is needed. In many cases it will be sufficient to minimise the existence of 'earth loops' by arranging all grounded leads in the conventional 'tree' structure, avoiding multiple paths between any two points. It has been generally found best to use the metal top plate of the cryostat as the 'trunk' of the tree. The braid used for the outer conductor in room temperature leads should be of very closely woven copper wire, again with any joins being soldered. It is vital to use coaxial leads everywhere. If these measures are insufficient and interference at mains frequency or harmonics of it are still a problem, another possible approach is to run all available instruments from battery power supplies. Fortunately this is possible with most general purpose laboratory instruments. Heavy duty car batteries provide a relatively cheap source of rechargeable power, but must themselves be well shielded inside a thick-walled conducting box. In such circumstances it is probably also necessary to enclose the cryostat and battery-powered devices in a Faraday cage (see section 7.1), coupling signals to external mains-powered and digital equipment via optical links.

7.5 A Simple, Adjustable Point-contact Josephson Junction

The reader may have gained the impression from the description of sophisticated junction fabrication techniques outlined in the previous chapter that the production of a Josephson junction is inevitably very complex. To dispel this notion we describe here a device which may be very simply constructed, which can be operated in a liquid helium storage dewar and which can demonstrate induced constant-voltage steps with the minimum of required microwave power. A device of this kind might form a cheap straightforward basis for a student demonstration of the Josephson effects.

7.5.1 Details of construction

The weak link is formed in the region where two interlocking superconducting wire loops touch. The wire used is phosphor bronze (of about 0.5 mm∅) which is coated with Pb–Sn solder, an alloy with a T_c of greater than 7 K. The critical current of the device may be adjusted over a wide range by adjusting the tension in the wire loops. One loop is attached to an insulating block which can slide in the vertical direction when moved by a coarse screw drive. With a conventional point-contact junction the mechanical drive must be very carefully designed and constructed. In this case it is unnecessary because the wires are wound into coil springs between the junction area and the terminal blocks. (For construction details see figure 7.3). The diagram shows the junction mounted in a length of *Q*-band waveguide (this must of course be made of low thermal conductivity material such as stainless steel) which forms the main support for the dewar insert and may also be used to direct applied microwaves at the Josephson junction. The lower end of the waveguide can be surrounded by a simple superconducting shield, which could be made from lead foil, soldered along the seams. Four fine copper wires twisted together connect the junction to current source and voltmeter at room temperature, to measure the current–voltage characteristic.

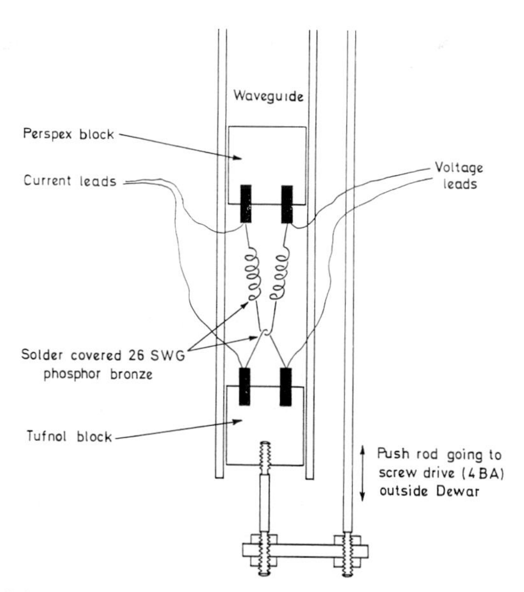

Figure 7.3 Schematic of a 'hair-pin' adjustable point-contact Josephson junction

7.5.2 Performance

Experience has shown that the adjustment mechanism allows fairly smooth variation of the junction critical current, from a few μA up to 100 mA or more. Provided i_c is less than about 10 mA, induced constant-voltage steps can be produced with as little as 1 mW of microwave power at the top of the probe. A solid state source such as a simple Gunn diode is suitable. The junction quality varies somewhat with the surface state of the solder-covered wires. It is believed that the influence of water vapour encourages the growth of 'whiskers' on the surface, producing microscopic points which form good weak links. The better junctions show induced steps of at least 50 μA wide at voltages up to 3 mV at 4.2 K. An audio-frequency oscillator can be used as a current source, applied to the x input of an oscilloscope. A simple differential input amplifier, based on a 741 op-amp for example, is adequate to display the step spacing. Figure 7.4 shows an I–V characteristic of a 'hair-pin' junction. The junction need not be mounted in a waveguide. Lower-frequency radiation (less than about 20 GHz) can be transmitted down coaxial cable with tolerable loss so that the junction could be located close to a loop termination of the coaxial cable.

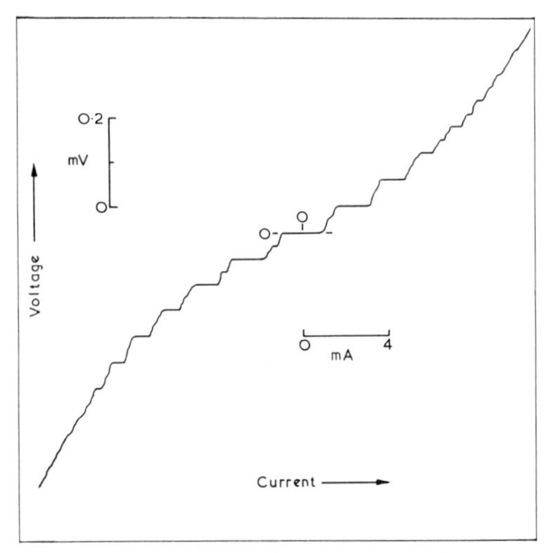

Figure 7.4 I–V characteristic for a 'hair-pin' junction when irradiated with 36 GHz microwaves

7.5.3 Possible projects

A junction of this type may be used in remarkably sophisticated applications of the Josephson effects. For some years such a device was the basis of the NPL Josephson voltage standard (Gallop and Petley 1974). It can be instructive to measure the induced step amplitudes as a function of applied microwave power. If the junction is situated in an almost closed cavity, self-resonant steps appear in the characteristic. In addition it is quite frequently found that double, or multiple, junctions are formed which act as DC SQUIDs. If the device is surrounded by a small solenoid a current flowing in this will modulate the parallel junctions' critical current. The number of demonstrations possible is only limited by the inventiveness of the experimenter.

7.6 Refrigeration

Almost all the devices dealt with in the previous chapters require a temperature in the liquid helium range for correct operation. At the present time this means that liquid helium itself is necessary at some point in the operating cycle. Two basic methods exist for achieving this temperature range. First one may take a dewar of liquid helium and immerse the cryogenic circuit in it, refilling the dewar from a bulk supply of liquid helium when necessary. This is relatively straightforward, though limited to small circuits and rather expensive in its use of liquid helium, which at the time of writing costs around £3 per litre. It can provide a rather stable but inflexible temperature of 4.2 K. For variable-temperature operation the apparatus must be incorporated in a 'cryostat', a vacuum-insulated vessel into which liquid helium may be introduced from a storage vessel via a vacuum-isolated transfer tube. When the cryostat is full the temperature of the liquid helium vessel, and that of the thermally linked superconducting device, may be reduced to as low as about 1 K by causing the liquid to boil under reduced pressure, and hence temperature, by pumping on the exhaust line from the vessel with a vacuum pump. A more rapid cool-down cycle, though less accurate temperature control, is achieved with a continuous flow cryostat, in which liquid helium is transferred continuously from the storage dewar into a small vessel containing the superconducting circuit. Consumption of liquid helium is also higher for this arrangement.

A second, fundamentally different, cooling technique involves attaching a helium liquefier to the electronic system, providing a continuous environment at 4.2 K or below. A variant of this is possible if a temperature no lower than about 8 K is required. Then a closed cycle helium refrigerator provides a simpler and smaller alternative to a liquefier, using only cold gas cooled by a two-stage Joule–Thomson expansion cycle. Continuous refrigeration, of either type, although growing rapidly in significance for large-scale superconducting installations in medicine, industry and high-energy physics, has hardly been used at all for Josephson devices. Several reasons underlie this reluctance. First the scale of superconducting electronics is at present small, both in physical and economic terms. The available helium refrigerators are large and expensive, better suited to producing many litres of liquid each day rather than a small fraction of a litre, which is all that a well designed SQUID or Josephson detector cryostat should require.

It would not be possible to produce a portable system, appropriate for a geophysical sensor for example, which included a refrigerator. This is mainly because a gas compressor is required to recycle the helium working fluid, after expansion through a Joule–Thomson (JT) valve. The compressors used are modified versions of machines designed for industrial purposes and are both large and power hungry, requiring cooling air or water.

7.6.1 Small-scale Stirling cycle refrigerators

An even more important reason why existing refrigerator systems are unacceptable is that they produce a great deal of magnetic noise. Rotating steel and soft iron components produce magnetic fields which are the dominant source of ambient magnetic noise over a range of several metres from the machine itself. A number of groups have been working on a solution to this problem. Stirling cycle refrigerators are very simple in principle. A stepped displacer moves the working gas from the hot region to the cold region, being compressed at the same time and getting heated. A heat exchanger cools the compressed gas before expansion through the JT valve. The simplicity of the device means that it can be built on a small scale. The work of Zimmerman has demonstrated that by making careful use of non-magnetic materials, such as ceramics and plastics, it is possible to reduce the noise level associated with the reciprocating motion of the displacer to a level where it is negligible compared with other sources of environmental noise. A prototype Stirling refrigerator has produced an operating temperature of 6 K and a SQUID gradiometer operating with this refrigerator has been able to measure the magneto-cardiogram of a subject in an unshielded environment (Zimmerman 1980). The total power consumption of the machine is only about 50 W, suggesting that it would be suitable for transportable SQUID or Josephson systems.

7.6.2 Micro-miniature refrigerators

An entirely different approach to the solution of the small superconductor refrigerator problem has been followed by Little (1987). He noted that the size of superconducting electronic circuits is small, and getting smaller with the development of thin film circuitry and miniaturisation. This led to the notion that it might be possible to use the same techniques of the semiconductor industry to produce a refrigerator of appropriate size. Using photolithography to define the pattern and chemical processes to etch gas flow channels, Little designed a combined counter-flow heat exchanger and JT expansion valve on a silicon substrate (see figure 7.5). The etched channel is covered by cementing a flat cover over the channel-bearing substrate. A high-pressure gas supply is attached to the inlet and the refrigerator operates, with no moving parts, for several hours from a small standard cylinder. Devices of this type are commercially available using nitrogen gas, which they are capable of liquefying, and are used quite widely to provide cooling for infrared detectors. In addition to simplicity and reliability the small scale of the system means that the time to cool to their operating temperature (say 80 K) is extremely short, as low as 2 s. These gas-powered refrigerators should be particularly attractive for cooling superconducting electronic devices made from the high-temperature superconductors which have transition temperatures significantly above 80 K, a temperature already easily attainable with commercial versions of the cooler.

Figure 7.5 Miniature Joule–Thomson refrigerator heat exchanger and expansion valve etched in Si

The inversion temperature of ^{4}He is only 25 K so it is not possible to use the JT effect to liquefy it starting from room temperature. It is necessary to use two other gases in a precooling stage to reach 25 K. In general nitrogen and hydrogen would be used. At the time of writing a prototype hydrogen liquefier has been produced and work is proceeding on a helium version. If this becomes a viable proposition the subject of cryoelectronics will be revolutionised. The sheer simplicity and speed of cooling down superconducting circuits should do a great deal to remove customer resistance and allow the unique properties of these devices to be widely exploited in all kinds of non-research-based areas.

7.6.3 Higher-temperature superconductors

We deal in more detail with the impact of high-temperature ceramic superconductors in Chapter 9. Prototype SQUIDs have already been operated at temperatures above the boiling point of liquid nitrogen (77 K). Although this is still some 200 K below room temperature the refrigeration problems associated with attaining such temperatures are vastly eased over what is required for conventional superconductors. In financial terms, liquid nitrogen is about 100 times cheaper than liquid helium. Its latent heat of vaporisation is some 300 times greater which means that simple foamed polystyrene buckets can be used to contain the liquid, rather than the expensive and complex evacuated double-walled cryostats and transfer tubes required for helium. The large latent heat and low cost mean also that the cooling procedure to 77 K can be much faster than that to the liquid helium temperature range.

8

Fundamental Physics with SQUIDs

The best energy sensitivity expected for a SQUID operating conventionally in the linear detector mode is believed, on both experimental and theoretical grounds, to be $\sim\hbar/2$ J Hz^{-1} (see section 4.3.1). As we have already stressed, in almost all real applications environmental noise actually limits the sensitivity at a level several orders of magnitude greater than this. However there are a number of fundamental experiments being carried on at the time of writing which are able to use the full capabilities of a quantum noise-limited SQUID. These investigations involve tests of general relativity, searches for fundamental particles and high-precision tests of some basic physical laws. In this section a variety of different experiments will be discussed in some detail, both because of the challenging nature of the effects which they are intended to investigate and because they provide classic examples of the type of high-precision measurement which is only possible using the unique properties of superconducting electronics.

The group at Stanford University, set up by the late Professor William Fairbank, has been outstanding in this area, carrying out a number of very fundamental tests of physical principles using SQUIDs and other superconducting electronics. Among these are (i) tests of parameterised post-Newtonian versions of general relativity, (ii) the search for free quarks in condensed matter samples, (iii) searches for free magnetic monopoles, (iv) measurement of the earth's gravitational force acting on single electrons and positrons, (v) attempts to measure the permanent electric dipole moment of the ^{3}He nucleus and finally (vi) development of the most sensitive displacement sensor in the world for use in a cryogenic gravity wave antenna.

In addition to the work of the Stanford group a number of other projects will be described which will serve to illustrate the power of the devices and techniques which are now available. We also attempt to explain in this chapter why superconducting electronics shows particular promise for investigations into fundamental physics, and speculate on some ultimate limits to measurement. Finally we deal at some length with the ideas which have recently come to the fore, and which may lead to novel devices in future: namely the description and development of a representation of weak superconductivity that is quantum mechanically conjugate to the treatment of the Josephson effects which have been detailed in the first seven chapters of this book.

8.1 General Relativity and SQUIDs

General relativity has existed as a theory for more than 60 years and until quite recently its experimental verification depended entirely on three astronomical measurements: the advance of the perihelion of the planet Mercury, the aberration of starlight grazing the surface of the sun, and the gravitational shift of spectral lines emitted by white dwarf stars.

Following the development of the Mössbauer effect in the 1960s it became possible to carry out the first laboratory-based tests of general relativity. The important point is that the relativistic correction to Newtonian gravitation is of order 1 in 10^{10} for a typical star with one solar mass, acting at the earth's surface, whereas even for a mass of 1000 kg the effect 1 m away will be only 1 in 10^{16}. However the Mössbauer effect is sensitive to frequency, and hence energy, shifts of 1 in 10^{17}. Confirmation of the theoretical gravitational red-shift of the emitted photons was obtained to 1 in 10^2. The unusual properties associated with superconductivity, which have formed the topic of this book, now join with the Mössbauer effect in allowing other laboratory tests of Einstein's general relativity. Proposals so far made include a test of the Lens–Thirring effect (that is the dragging of inertial frames by a rotating mass), to be carried out by observing the standing wave pattern in a very high-Q superconducting toroidal microwave cavity, which is being rotated along with a coaxial spinning mass of some 1000 kg. There are a number of tests of non-Einsteinean components of relativity which may be seen as searches either for non-metric components of the gravitational interaction or for other contributions of the mass–energy and momentum distribution to the curvature of space-time. Many different authors have proposed possible effects involving, for example, non-conservation of PT (parity inversion combined with time reversal) invariance by the gravitational interaction or that the interaction is anisotropic, having some dependence on the spatial distribution of mass in the galaxy. The Einstein equivalence principle (EEP) is a most powerful statement which allows any experimental set-up to be analysed so that a non-metric contribution to the theory of gravitation will show up as a violation of EEP. No such violations have yet appeared at the time of writing.

An interesting analysis of a number of possible experiments involving superconductivity has been given by Brady (1983), which demonstrates among other things the improbability of a Josephson effect gravity gradient interferometer. A rotational analogue of Mach's principle is also proposed, which might be examinable using an absolute rotation sensor, known by the acronym SARDINE. In addition a novel type of superconducting gravity wave detector is outlined. We ought to ask at this point 'why is it that superconductivity seems able to satisfy many of the extremely severe requirements of laboratory-based relativity experiments?' The first simplistic answer invokes the existence of SQUIDs, with their unparalleled sensitivity. But in addition lossless screening properties are also crucial to many measurement situations, in providing an electrically and mechanically quiet environment in which extremely small effects may be detectable. In addition high-Q resonators and other very low-dissipation systems, good temperature control and very low ageing rates at low temperatures all provide important inputs.

8.1.1 The superconducting gyroscope experiment

The Stanford University group's relativistic gyro project must be one of the most ambitious projects undertaken in low-temperature physics (Turneaure *et al* 1989) . Since the original proposal by Fairbank and Everitt a large number of seemingly intractable problems have been solved in a step-by-step approach towards completion of the project (Anderson *et al* 1982). The essential principle is that general relativity predicts that a gyroscope in earth orbit should exhibit two distinct, though minute, precessional motions. The first, analogous to the precession of the perihelion of Mercury, should amount to about 55 arc sec/yr. An even smaller effect should arise from the Lens–Thirring frame dragging by a rotator (in this case the earth itself) and this is predicted to be only 0.7 arc sec/yr (see figure 8.1). The precision rotor consists of a quartz sphere, 67 mm in diameter, spherical to ± 1 nm. Its axis of rotation must be most carefully measured with respect to some fixed axis defined relative to the positions of distant stars. The rotor's homogeneity is of prime importance, to such an extent that any fiducial mark would cause an imbalance sufficient to obscure the expected effects of general relativity. Instead a unique property of superconductors is exploited to avoid the need for any marks. The surface of the sphere is coated with a thin uniform film of niobium. It is levitated in a quartz housing by means of electrostatic forces and then 'spun-up' by jets of helium gas. A rotating superconductor generates a magnetic moment (the London moment, see Appendix A.3.2). Very small changes in the flux linking a thin film primary of a flux transformer, associated with changes in the direction of the London moment, are sensed by a SQUID. The single-turn primary coil is evaporated on the inner surface of the quartz rotor housing. Detection of tilts of the predicted amount requires a SQUID with a sensitivity close to the quantum limit. Many other problems associated with this sophisticated space-borne experiment have had to be solved, including notably sustaining refrigeration of the system to 2 K for the planned flight time of two years, and production of a very accurate spacecraft orientation system based on a telescope locked onto a fixed star. Problems with the space shuttle programme have delayed the launch of the gyro experiment which is not now expected to fly before 1991.

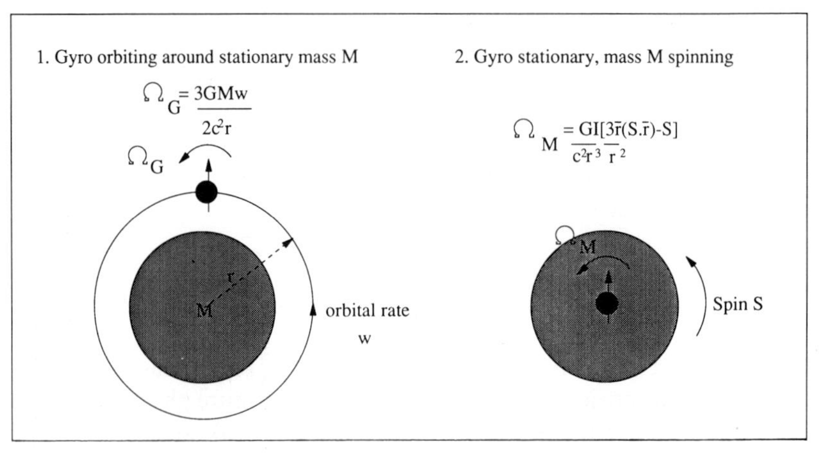

Figure 8.1 Two effects on the precession rate of a gyroscope in the neighbourhood of a massive body

8.1.2 Hughes–Drever experiment employing superconductivity

It is a direct result of EEP that all atomic clocks, of whatever construction, will run at the same rate if situated at the same point in space-time. The universal gravitational red-shift formula turns out to be a property of any self-consistent metric theory of gravity. Hughes *et al* (1960) and Drever (1961) carried out independently the first high-precision tests of EEP by observing the free precession frequency of a system of ^{7}Li nuclei, situated in a magnetic field which was maintained as constant as possible. This amounts to a comparison of a quartz crystal clock (i.e. the time base of the free precession frequency counter) and a nuclear magnetic resonance clock. The two are compared as the rotation of the earth changes the gravitational potential in which the clocks find themselves. The Hughes–Drever result, showing that the clocks ran at the same rate with no 24 hour periodic variations at the level of 1 in 10^{20} has been described as the most sensitive experiment in the whole of physics. This is because the energies involved in the states of the ^{7}Li nucleus are huge compared with the minute splitting of the Zeeman levels of the nuclear spin system which the free precession process mixes together.

There are a number of ways of interpreting the Hughes–Drever result (see for example Will 1981) but all authors are agreed that it provides an exceptional test of EEP. The importance of superconductivity to this result is that the superior magnetic shielding property of a superconducting tube allows the magnetic field in which the nuclei find themselves to be kept constant to a higher level of precision than was possible with the conventional electromagnets used by the original experimenters. Furthermore a better signal-to-noise ratio is available by using a SQUID to detect the free precession signal (see section 5.10). Gallop and Petley (1983) have already shown, with some very preliminary experiments, that a further factor of 1000 is easily attainable, and work to date has far from exhausted the potential for improving 'the most sensitive experiment in physics'.

8.2 Gravity Wave Detection

According to the theory of general relativity the distribution of mass–energy, momentum and momentum flux in a nearby volume determines the curvature of space-time at any point. Any change in the mass–energy distribution produces a change in curvature, which will propagate away from its origin at the velocity of light. There are analogies with the propagation of a displacement in a stiff solid such as steel, except that the stiffness is greater in the case of space by a factor of around 10^{30}. The supposed existence of a travelling disturbance in the local metric of space-time amounts to stating that 'gravity waves' can exist, and can transmit energy. At the time of writing there has been no direct detection on the earth of the energy associated with gravity waves, although a number of searches have been going on for several years. Perhaps surprisingly this is not because the amount of energy generated in the form of gravity waves is so small, at least for the case of cosmic-scale events. Rather it is that the coupling between this energy and ordinary amounts of matter, such as can be manipulated in a laboratory, is exceedingly weak.

Any attempt both to generate and to detect gravity waves in a laboratory is far beyond the capabilities of today's technologies. For example, for two masses M_1 and M_2 orbiting each other with a separation of $2r$ and period T the power radiated is

$$\{2^{11}\pi^6 G r^4 [M_1 M_2/(M_1 + M_2)]^2\}/5c^5 T^6 \tag{8.1}$$

where G is the universal gravitational constant. Thus for two 1000 kg masses 10 m apart rotating around one another at 1000 RPM the radiated power in gravity waves is only 10^{-30} W. On the other hand for a binary star system, each member being of 10 solar masses, with a separation of 0.1 light years and an orbital period of 1 year, the radiation flux at the earth, 1000 light years from the system, would be $10^{-11}\ \mathrm{W\,m^{-2}}$, an apparently reasonable flux, easily detected if it were in the form of electromagnetic radiation.

In order to maximise the coupling of gravitational radiation to matter, two approaches are possible. One is to employ an antenna in the form of the largest mass available, while another uses two smaller masses separated by a very long baseline. At present both approaches are being followed and SQUIDs have only found uses in the first technique. The second method is being developed in the form of a giant multi-pass optical interferometer. It is not clear at the time of writing which of the two antenna types will prove most sensitive.

8.2.1 SQUID displacement transducer

The essential idea behind this device, and its subsequent development, came, like so much else in this chapter, from Professor Fairbank's group at Stanford University (Paik 1980). A thin high-purity niobium diaphragm is mechanically coupled to a massive aluminium bar, the combination being maintained at a stable cryogenic temperature (to reduce the effects of Nyquist noise) whilst being extremely carefully isolated mechanically and electromagnetically from its surroundings. The massive bar acts as the antenna for gravitational radiation because an incident quadrupolar wave will cause alternate elongations and compressions of the bar, with a corresponding antiphase motion in the direction perpendicular to its length. The high purity of the bar and the low acoustical loss intrinsic to aluminium ensures that it will possess a number of very high-Q resonant modes and consequently that an incident disturbance will excite one or more of these modes into oscillation. The mechanical parameters of the diaphragm are chosen so that it has a resonance close in frequency to one of the bar's modes of oscillation and the two act as weakly coupled oscillators.

The diaphragm has two flat spiral ('pancake') coils of superconducting wire mounted on either side of it, wound in series opposition. These act as the primary of a flux transformer, the matched secondary of which is coupled to a very low-noise DC SQUID. The complete flux transformer circuit carries a circulating supercurrent. A small movement of the flux-excluding diaphragm towards one coil and away from the other, such as would occur if the antenna was set into oscillation, will change the effective inductance of each coil, causing the net circulating supercurrent to change in order to maintain the magnetic flux linking the transformer constant. This current change is then detected as a flux change at the SQUID. Figure 8.2 shows a schematic view of the mechanical construction of the resonant displacement transducer, together with the electrical equivalent circuit of the coils. The mechanical self-resonant frequency, primarily determined by the stiffness of the diaphragm, has a magnetic component which varies with the amount of trapped magnetic flux linking the pancake coils, so that adjustment of the current allows the resonant frequencies of antenna and transducer to be brought arbitrarily close together. In the case of the 4500 kg bar at Stanford the resonant mode selected is at around 800 Hz. Ambitious plans called originally for the whole system to be cooled to the millikelvin region, with the antenna levitated above a superconducting persistent current coil to provide mechanical isolation. At the time of writing, a mechanical suspension is used and the bar is operated at 1.5 K. For a recent review of resonant bar gravity wave antenna see the review article by Blair (1987).

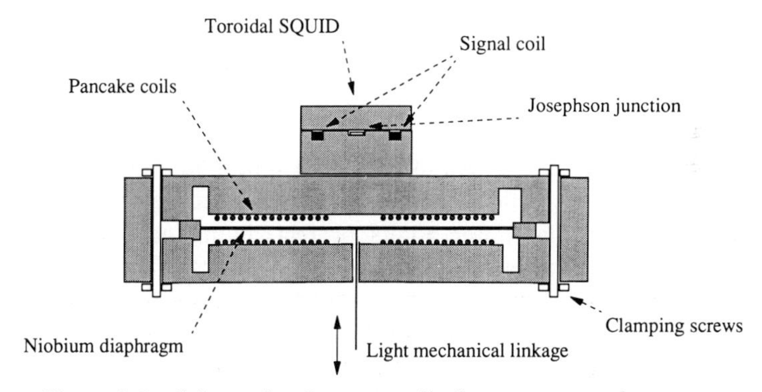

Figure 8.2 Schematic of a SQUID displacement transducer

The thin film DC SQUID used is of the type developed by IBM (see section 4.4) which operates close to the quantum limit (Awschalom *et al* 1989). The total system has demonstrated a displacement sensitivity of better than 10^{-19} m in a bandwidth of 1 Hz, corresponding to 1 in 10^4 of the diameter of a typical nucleus. This should be adequate to detect the radiation emitted by a supernova within our galaxy or one of its near neighbours, an event which may be expected about once every 30 years. The antenna in its present form detects many events each day and these are attributed to environmental disturbances including microseisms as well as man-made mechanical noise. To eliminate these local effects it is necessary to run two or more similar, but widely separated, detectors in co-incidence mode. A set of three detectors, in California, Louisiana and Rome, has been running for some time now. Ironically supernova 1987A, in the lesser Magellanic cloud, should have produced sufficient signal to be detected but at the crucial time all three detectors were out of action! Only a number of 'first-generation' room temperature bar antenna systems were operational, some of which may have indicated the passage of gravity waves, although this must be regarded with some doubt at present.

8.3 Detection of Fundamental Particles using SQUIDs

Two more experiments carried out at Stanford University by the same group who built the cryogenic gravity wave detector described in the previous section have caused considerable interest in the world of high-energy physics. The first of these was a search for free quarks (Fairbank 1977), the second being the tentative observation of a magnetic monopole (Cabrera 1982).

8.3.1 Search for free quarks

The apparatus used in this search is a rather sophisticated cryogenic version of the classic Millikan oil-drop experiment. Here the oil drops have been replaced by very small superconducting niobium spheres which can be levitated in a combination of magnetic and RF electric fields (Fairbank 1977). If its mass is known the acceleration of the sphere in a known field allows the electric charge to be deduced. A SQUID is used to detect the position of the levitated ball which, due to its perfect diamagnetism, perturbs the inductance of a coil system coupled to the secondary of a flux transformer. The balls typically carry spurious electric charges acquired from the environment, but these may be eliminated by a process of trial and error by bringing up electron or positron emitters close to the ball until the charge on it leaks away and reaches a minimum. For a number of the balls used, the residual charge was not zero but close to $e/3$, where e is the electronic charge. This is just the value to be expected if a single free quark existed on a ball. However the results are controversial in that searches using other materials with equal or greater sensitivity have failed to observe such residual charges on a total mass of material much greater than that examined by the Stanford group.

8.3.2 Magnetic monopole detection

In 1932 Dirac, beginning from some simple arguments based on the gauge invariance of electromagnetism, predicted that the existence of a single magnetic monopole anywhere in the universe, with charge h/e, was sufficient to explain the observed quantisation of electric charge in units of e. (Ironically, if one believes the result of the previous section, that the fundamental unit of free charge is that associated with a quark ($e/3$), then the monopole charge should be a factor of three greater: $3h/e$.) More recently gauge theorists and cosmologists have come together to suggest that magnetic monopoles should exist if Grand Unified Theories of matter are correct. The inflationary model of the early history of the universe supports this view, although the number of monopoles required in any universe may be as small as one!

The contribution of superconductivity to magnetic monopole searches arises from the distinctive signature that passage of a monopole through a superconducting loop would supply. A simple application of one of Maxwell's equations, generalised for the existence of a magnetic monopole current density $\boldsymbol{j}_{\mathrm{m}}$

$$\nabla \times \boldsymbol{E} = -\,\mathrm{d}\boldsymbol{B}/\mathrm{d}t + \boldsymbol{j}_{\mathrm{m}} \tag{8.2}$$

shows that the result of the passage is to increase the magnetic flux linking the ring by $h/e = 2\Phi_0$. This would provide a rather unambiguous indication since the passage of a magnetic dipole or an electrically charged particle will only cause a transient pulse in the flux through the ring, whereas a monopole will cause a step change which will persist as long as the ring remains superconducting. A SQUID magnetically coupled to the ring would be easily capable of detecting such a change.

A detector of this type was built by Cabrera (1982) at Stanford and was run initially for a period of about 150 days. During this time a single event of the correct magnitude was observed. Indeed there were no other events which were at all close to the right magnitude. In view of subsequent searches using much larger loops for longer periods it does not now seem likely that this observation was of a monopole passage. There are some proposed though improbable processes which could have mimicked the event, such as an internal stress relief event which had sufficient energy to liberate simultaneously two trapped fluxons from the body of the loop to its interior. A process of this type would have been particularly unlikely for the experimental arrangement used by Cabrera since the loop and its associated supports and the SQUID had been cooled down in a very low magnetic field (<10 pT), provided by a special expanded lead balloon superconducting shield.

To increase the 'cross section' of the detector one might simply expect to have to increase the area A of the superconducting loop. This straightforward approach has two disadvantages. Firstly, the ambient field change required to simulate the passage of a monopole, $\delta B = 2\Phi_0/A$, falls as A increases so that screening from external field changes becomes more critical. Secondly, the inductance of the ring increases approximately linearly with A so that it becomes increasingly difficult to match a large coil with the μH inductance of the SQUID input coil (see section 5.2). Instead of a simple single coil a complicated gradiometer winding (see section 5.3 and figure 8.3(b)) has been employed, in which multi-loop planar coils are arranged so that adjacent loops have opposite winding senses, which both reduces the sensitivity of coupling to external fields and the total coil inductance. Several such coils, each coupled to its own SQUID, have been run in coincidence in the same cryostat with a number of other detectors

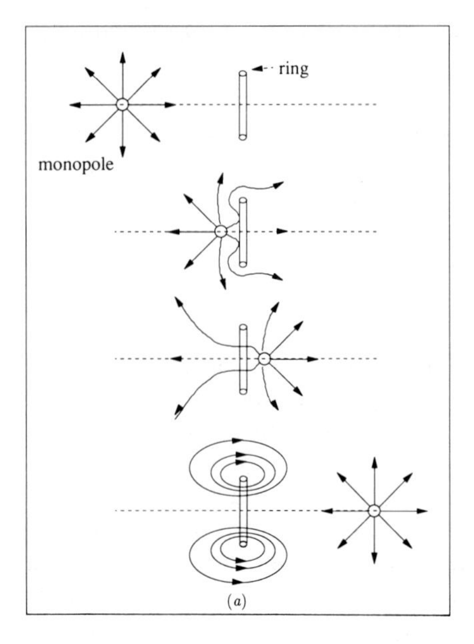

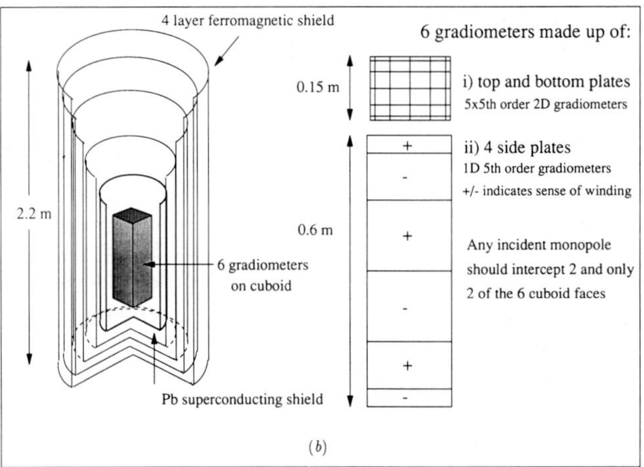

Figure 8.3 (*a*) Schematic of the passage of a monopole through a superconducting ring, and (*b*) schematic of a 0.1 m^2 area monopole detector (Bermon *et al* 1985)

of electromagnetic or mechanical transients which might mimic the passage of a monopole (Bermon *et al* 1985). Groups at IBM, Fermilab and Imperial College, as well as Stanford University, have now run quite large detectors for periods of hundreds of days at a time. Only one extra very tentative event has been observed (Schouten *et al* 1987) but the additional accumulation of data has pushed the upper limit on the monopole flux down to

$2.2 \times 10^{-13}\ \mathrm{cm^{-2}\,sr^{-1}\,s^{-1}}$, some four orders of magnitude lower than Cabrera's initial estimate. Nevertheless the search continues. Designs have been drawn up for an extensive 'cryostat farm' of detectors which would have a total effective collecting area of $100\ \mathrm{m}^2$. This project may never be built, but the interest in the existence of magnetic monopoles (or lack of it) seems certain to persist for many years.

8.3.3 Dark matter searches using SQUIDs and Josephson junctions

Cosmologists have been worried for a number of years about an insufficiency of observed mass in our galaxy and others to provide for the observed motions of stars about the galactic centres. This has led to the proposal that perhaps 90% of the galactic mass is in some 'dark' form. Weakly interacting massive particles (WIMPs) are favoured by many at present and a number of experiments are being set up to try to detect these elusive particles. The article by Dekel and Rees (1987) summarises some of the search techniques being used. Superconductivity in the form of Josephson or quasi-particle tunnel junctions is playing an important role in providing very sensitive photon detectors and discriminators (for example, resolution of 1 kV x-rays with a sensitivity of only a few eV is possible).

8.4 The Josephson Voltage–frequency Relationship

The first real application of the Josephson effects was to the measurement of the fundamental constant $h/2e$ and later to the realisation of a quantum standard of voltage based on the AC effects. The earliest versions of this superconducting voltage standard consisted in essence of one or at most a few junctions connected in series, giving induced constant-voltage steps extending to a few mV for each junction. (Large-amplitude steps can only be induced by microwaves up to a voltage of order of the superconducting energy gap, $\sim 2\Delta/e \sim 5$ mV.) To relate the quantised voltage to useful transfer standards such as Weston standard cells (EMF of 1.018 ... V) highly accurate and complex potentiometry is required. Although SQUIDs and direct current comparator ratio devices (see section 5.5) make this possible with high accuracy, operation of these systems remains a highly skilled art. The vastly simpler systems based on large numbers of identical series-connected junctions which are now becoming available give both improved accuracy and much simplified operation and several novel extended applications have been proposed. These junction arrays are dealt with in a following section.

Customers demand ever more accurate precision from calibration services and although the present level of confidence (<10 nV or 1 in 10^8)

is adequate, further advances will no doubt be demanded. It is not possible to check the accuracy of equation (2.11) against any other more stable voltage, since it was the lack of such a reference which brought the Josephson voltage standard into existence. The best that can be done is to test the independence of the relationship by changing various parameters such as temperature, the material from which the junction electrodes are made, magnetic field, geometry etc. The most sensitive of these tests has involved comparing the voltages appearing across two dissimilar junctions when they are irradiated with the same microwave frequency. The first experiment of this kind, due to Clarke (1972), achieved high accuracy even though the voltage bias of each junction was very small. This arose because the voltage difference, if present, could be measured with extremely high accuracy using a SQUID. This is an ideal application for a quantum interference device since the junction input impedance when current biased onto a constant voltage step is close to zero, which optimises the voltage sensitivity of the SQUID. Various modifications of this basic arrangement have been reported (MacFarlane 1973, Gallop 1973), the latter dispensing with a separate SQUID and using two irradiated junctions in a superconducting low-inductance loop as a form of DC SQUID. The microwave-induced step width is modulated by any contribution to the circulating supercurrent flowing around the loop, which would be generated by a small voltage difference between the two junctions. If this difference is δV the circulating supercurrent i grows linearly with time according to the expression

$$\mathrm{d}i/\mathrm{d}t = \delta V/L \qquad (8.3)$$

where L is the loop inductance. Minimising L allows very high sensitivities to be achieved. The best precision so far reported (Jain *et al* 1987) arises from a conventional measurement of this type which demonstrated that $\delta V < 2 \times 10^{-21}$ V between a Nb–Cu–Nb junction and a microbridge of In when both were irradiated with 90 GHz microwaves and biased on the 4th step. This suggests that equation (2.11) is independent of material to better than 1 in 10^{16}.

The same group has refined this technique to demonstrate the Einstein Equivalence Principle. Two identical Josephson tunnel junctions are separated vertically by a distance of 73 mm and irradiated with the same phase-locked microwave source (see figure 8.4). The gravitational blue-shift between the photons reaching the lower junction compared with those being absorbed by the higher one might be expected to produce a higher voltage drop across it. However this is exactly compensated by a blue-shift in the energy of a charged particle (a superconducting pair, for example) in being transferred from the upper to the lower junction. A SQUID magnetically coupled to the two-junction circuit with each biased on the same order Shapiro step showed no circulating current between them. This set an upper limit on the voltage difference between them of $< 10^{-22}$ V, testing the Einstein Equivalence Principle to $\pm 5\%$ (Jain *et al* 1987).

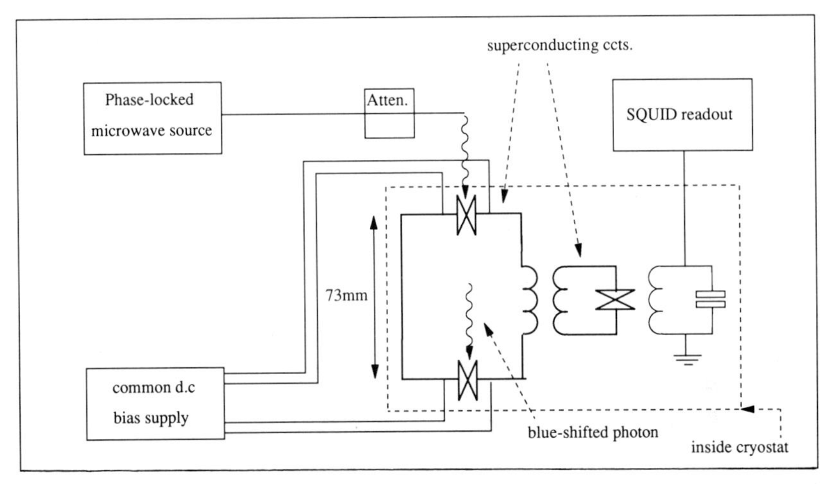

Figure 8.4 Circuit used to measure gravitational red-shift of charged particles (Jain *et al* 1987)

8.4.1 An unbiased junction array voltage standard

A recent development has already made obsolete the complex potentiometry of the first generation of Josephson voltage standards, by using an array of more than 1000 nearly identical series-connected junctions. Although in principle such a series array could consist of individually current-biased junctions sufficient to produce an output voltage of order 1 V, this would prove extremely difficult to operate in practice. Levinsen *et al* (1977) proposed an ingenious way to circumvent this complexity. They pointed out that a microwave-irradiated junction with sufficiently large capacitance C such that the hysteresis parameter $\beta_c = 2ei_cR^2C/\hbar > 1$ can exist with a finite voltage across it and the order parameter phase difference evolving in step with the phase of the applied electromagnetic radiation but with no bias current. (This behaviour can be simply understood in terms of the tilted washboard mechanical analogue of the Josephson effects, treated in some detail in Appendix B.1.) For the limiting condition that β_c becomes exceedingly large the microwave field looks more and more like a true voltage source to the junction and the induced nth-order step amplitude varies like the nth-order Bessel function. Then the nth step will cross the zero current axis if $n\hbar\omega/2ei_cR < J_n(\max)$. An I–V characteristic for such a junction is shown in figure 8.5.

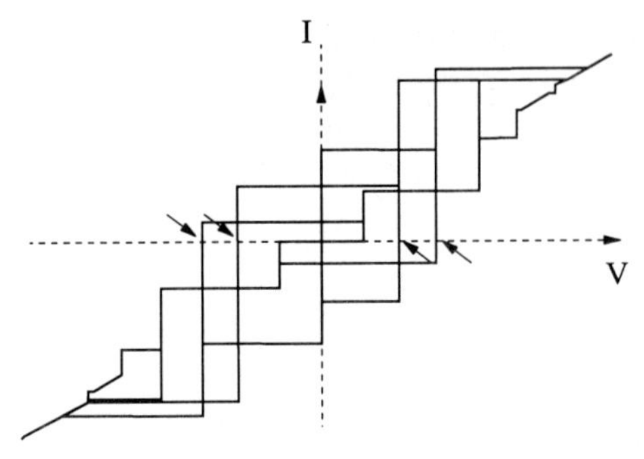

Figure 8.5 Microwave-induced zero-crossing constant-voltage steps in a single Josephson junction

Experimentally, finite capacitance junctions also exhibit current axis-crossing induced microwave steps. Levinsen *et al* (1977) proposed that it might be possible to excite a series array of axis-crossing junctions to give a quantised voltage close to 1 V, but it was left to Niemeyer *et al* (1985) to demonstrate how this might be done. A crucial problem which had to be solved was how to arrange that each junction experienced essentially the same microwave field strength so that equivalent induced steps would have approximately the same current width greater than some minimum level which would provide stable operation. The solution was to incorporate the junctions (Pb–In counter electrodes with oxide barriers) as the upper electrode in a superconducting transmission line (with an Nb counter electrode), carefully terminated with a matched load so that there was no reflected power (see figure 8.6). The array was irradiated with 90 GHz radiation. Current bias is only necessary to set the voltage approximately across the array. (In the limit there are two apparently degenerate conditions of the array, with equal but opposite voltages across it so that an initial

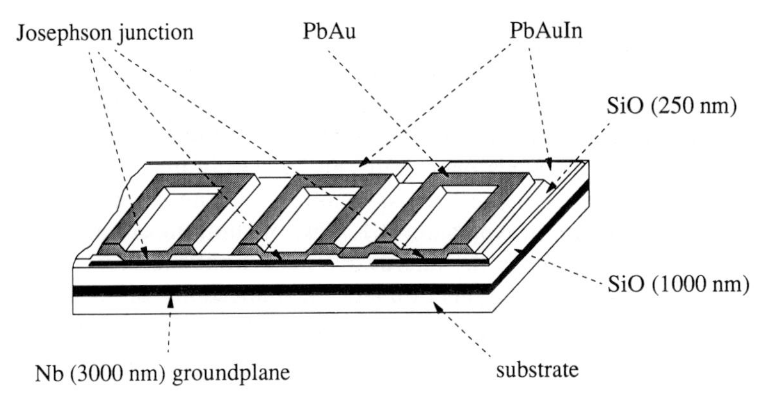

Figure 8.6 Section of Josephson junction voltage standard array

Figure 8.7 An array of Josephson junctions, arranged as a meandering microwave stripline, used to maintain the NPL representation of the volt. The device and this photograph were provided by courtesy of NBS, now renamed as NIST, of the USA

current bias pulse is necessary to lift the degeneracy.) Figure 8.7 shows an overall view of such an array made at NIST in Boulder. At the time of writing it has not yet proved possible to produce an all-refractory array but 1 V junctions using 1474 junctions and even a 10 V array with 19 000 junctions are in service as extraordinarily accurate and convenient voltage standards (see Niemeyer *et al* 1985, Hamilton *et al* 1989 and Kautz *et al* 1987). These devices will undoubtedly revolutionise potentiometry, possessing accuracy at the 1 in 10^{10} level, without recourse to calibration by standards laboratories.

8.5 The Aharonov–Bohm Effect

A classical result of low-energy atomic physics, discovered some 40 years ago, is the Aharonov–Bohm (AB) effect, which demonstrated convincingly the physical reality of the electromagnetic potentials. Since at least the time of Maxwell it had been assumed that the 'real' electromagnetic variables are electric and magnetic fields $\boldsymbol{E}$ and $\boldsymbol{B}$. That $\boldsymbol{E}$ and $\boldsymbol{B}$ could be derived from a scalar electric potential φ and a vector potential $\boldsymbol{A}$ respectively was viewed as merely a happy accident. The lack of reality of the potentials was thought to be demonstrated by the existence of the property of gauge invariance (see Appendix A.3). Any arbitrary potential function $\chi(r, t)$ can be added to φ and $\boldsymbol{A}$ provided that it satisfies the following conditions:

$$\varphi' = \varphi + \mathrm{d}\chi/\mathrm{d}t \tag{8.4}$$

$$\boldsymbol{A}' = \boldsymbol{A} + \nabla\chi \tag{8.5}$$

without in any way apparently affecting the physical results of any measurement. Aharonov and Bohm suggested a straightforward test of this assumption. They proposed that a two-slit diffraction experiment be performed on an electron beam, with a long thin solenoidal coil being placed between the slits and immediately behind them (see figure 8.8). The experimental question to be answered is then 'what effect will a direct electric current through the solenoid produce on the electron diffraction pattern?' A direct application of conventional quantum mechanics led Aharonov and Bohm to suggest that the wave function of an electron will be multiplied by a term $\exp(\mathrm{i}q/\hbar \int \boldsymbol{A} \cdot \mathrm{d}\boldsymbol{s})$ in going from the point a to point b, where $\boldsymbol{A}$ is the magnetic vector potential at each point on the path and q is the electric charge on the particles in question. The effect of such a term on the diffraction pattern arises from the sum of two such phase shifts on the electron wave functions, from integration paths on either side of the solenoid. Then the total phase shift becomes:

$$\delta\varphi = q/\hbar\left(\int \boldsymbol{A} \cdot \mathrm{d}\boldsymbol{s} - \int \boldsymbol{A} \cdot \mathrm{d}\boldsymbol{s}\right) = q/\hbar \oint \boldsymbol{A} \cdot \mathrm{d}\boldsymbol{s} \qquad (8.6)$$

where the final integral is performed around a closed path enclosing the solenoid. From Stokes' theorem and the relationship between magnetic vector potential and flux density $\boldsymbol{B} = \nabla \times \boldsymbol{A}$, the integral is just equal to the magnetic flux linking the path which is generated by the solenoid. However since the solenoid is very long the flux density $\boldsymbol{B}$ everywhere outside it is negligibly small. Thus the electron wave functions (which are assumed to have negligible amplitude on the surface and in the interior of the solenoid) would experience a negligible phase shift if direct experience of the field quantities themselves were necessary, and the potential functions were merely a mathematical convenience. (The decay of the wave function at the surface of a superconductor takes place exponentially on the length scale $\hbar/(2mW)^{1/2}$ where W is the work function. This is typically only ~ 1 pm, so on any reasonable scale of fabrication the order parameter amplitude at the solenoid is quite negligible.)

The first experiment was carried out by Chambers (1960), who demonstrated the essential correctness of the AB analysis. The electrons did produce the phase shift predicted and the reality of the potentials in quantum mechanics was confirmed. Not long afterwards an even more convincing demonstration of the AB effect was made using a SQUID by Silver *et al* (1967).

The discussion outlined above, and the experimental results, demonstrate that it is not possible to add an arbitrary position-dependent scalar function $\chi(\boldsymbol{r})$ to the potentials, even when it satisfies the above conditions, without producing detectable physical effects (in this case a lateral shift of a diffraction pattern as the applied field is changed). In other words local gauge invariance no longer is a good symmetry (see Appendix A.3). Global gauge invariance remains a good symmetry since it is equivalent to adding an arbitrary but constant amount to the phase of the wave function at each point in space, and this will have no detectable physical consequence.

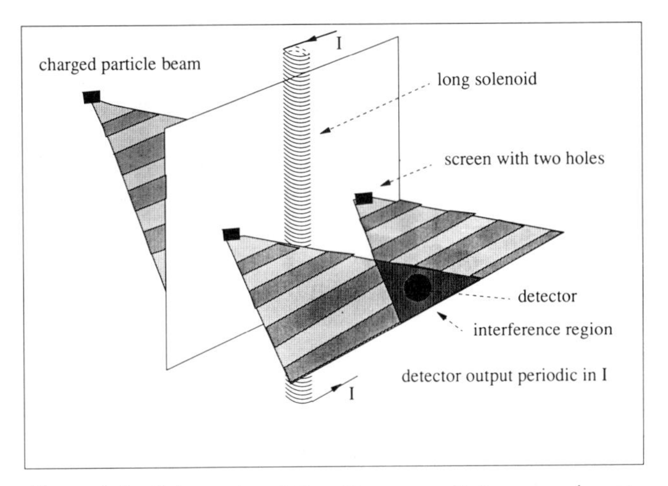

Figure 8.8 Schematic of the Aharonov–Bohm experiment

8.6 A Fully Quantum Mechanical Model of Weak Superconductivity

Josephson was led to the prediction of the effects that bear his name from a consideration of the coupling of the order parameters ψ_1 and ψ_2 in two pieces of superconductor as the coupling between them is varied. Thus for complete separation the order parameters and hence their phases are completely independent. At the other extreme when the two pieces are in intimate physical contact the combination is effectively a single piece of superconductor, so that the phases are rigidly coupled. Josephson argued that there is an interesting intermediate coupling situation in which the phases of the order parameters are only weakly coupled and the resulting phase difference can be smoothly varied under the influence of a number of external parameters such as current, voltage and magnetic field.

The picture which Josephson adopted required that the order parameter of the superconductor should possess everywhere a well defined phase, or equivalently if multiply connected, a well defined flux state. Such a representation is appropriate to deriving the Josephson effects, but it was pointed out by Anderson early on in the development of the subject that an alternative description of the state in terms of a well defined Cooper pair occupation number N is also possible. Anderson (1964) showed that the phase φ and pair charge excess Q are quantum mechanical conjugate variables, connected by the commutation relationship

$$[Q, \varphi] = -2ie. \tag{8.7}$$

A state with a well defined phase can be described by a linear combination of states corresponding to different values of Q:

$$\psi(\varphi) = \sum_{Q} a(Q)\exp(i\pi Q\varphi/e)\psi(Q).$$

It is clear from this argument that there is an alternative representation in terms of a state with well defined Q, made from a linear combination of different phases:

$$\psi(Q) = \sum_{\varphi} a(\varphi)\exp(iQ\varphi/2e)\psi(\varphi).$$

Such a state was only mentioned in passing by Anderson and received no further attention. However more recently the possibility of macroscopic quantum tunnelling in weak superconducting circuits has caused a great deal of interest in this state. Consider a singly connected weak link, whose free energy $U(\varphi)$ is represented by the 'washboard' potential, described in Appendix B.1 (see figure B.1). Leggett (1980) proposed that the dominant decay mode from a metastable state in the tilted washboard, corresponding to a current-biased junction, would be by tunnelling if the condition $kT \ll \hbar\omega_0$ could be satisfied, where ω_0 is the frequency of small oscillations about the energy meta-minimum.

Perhaps surprisingly in view of the acronym, until this time (around 1981) SQUIDs had always been regarded as semi-classical devices, exhibiting quantised flux but being describable by classical quantities such as voltage and current. The success of this unrealistic approximation is believed to result from the fact that the devices had always been used in temperature or noise regimes where flux transitions are dominantly driven by thermal fluctuations or external noise sources. A number of groups have reported experimental observations of such tunnelling processes in tunnel junctions and in SQUID rings. The observed tunnelling rates are in reasonable agreement with WKB calculations performed by Caldeira and Leggett (1981), for single junctions, and resonant activation under microwave irradiation has demonstrated the existence of quantised levels within the periodic washboard potential (figure 8.9(*a*), Martinis *et al* 1987). Clark and co-workers (1981–6) at Sussex University have dominated the experimental field concerning SQUID rings, demonstrating both energy band structure in flux-periodic and charge-periodic structures. The transition from one representation to the other occurs as the temperature is lowered from about 4 K to 2 K.

Theory and experiment cannot yet be said to be in unanimous agreement. There are two outstanding problems. First it is not clear how to deal with dissipation in macroscopic systems, where the coupling to the environment is extremely complex. Secondly the appropriate treatment of both electromagnetic bias sources and of detectors is unclear. That a true quantum mechanical treatment of Josephson and SQUID devices is required may be seen from figure 8.9(*b*) which shows that the uncertainty principle limit for a linear amplifier is closely approached by the best superconducting devices over a wide frequency range.

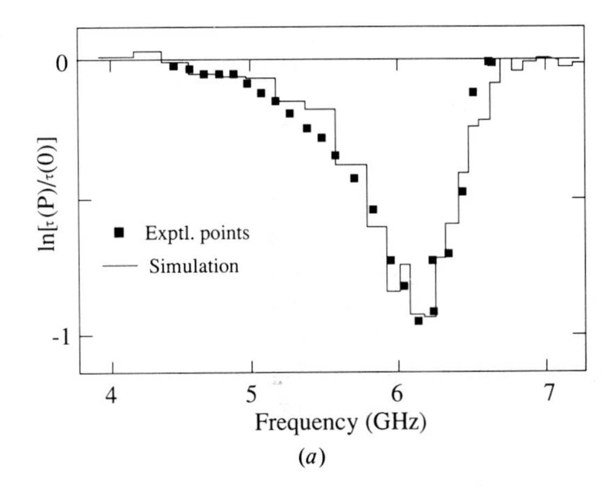

(*a*)

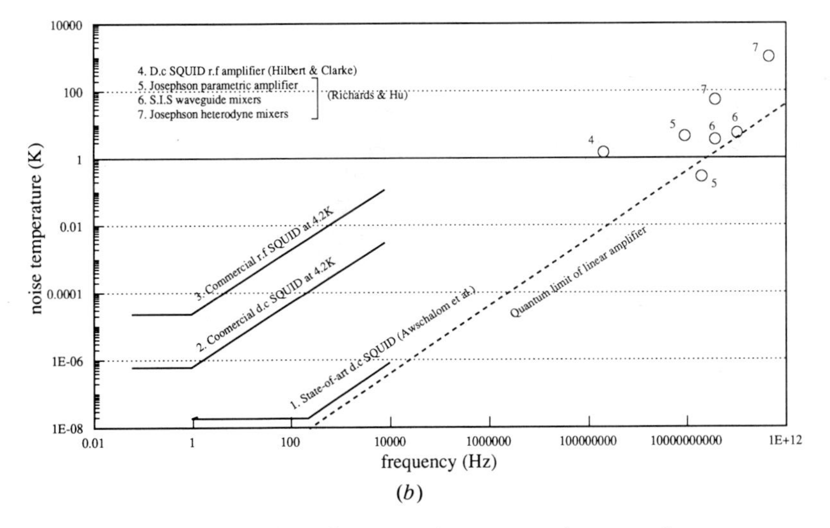

(*b*)

Figure 8.9 (*a*) Resonance in escape time versus microwave frequency. The vertical axis is the ratio of the lifetime of the junction in a zero-voltage state with and without applied microwave power, P, at frequency f. (*b*) Noise temperatures for various superconducting amplifiers as a function of frequency

By way of speculation, encouraged by the general lack of agreement, we shall continue to argue by analogy to suggest that there may, in addition to the basic flux tunnelling possibility, be further important and useful properties associated with this new quantum mechanical view of SQUIDs. First it is fairly clear that the Anderson view that the order parameter may have two possible representations based on fixed phase or fixed pair occupation number has been confirmed by the theory and experiments of the Sussex University group, the number–phase commutation relationship being equivalent to charge and flux being canonically conjugate variables. Thus for a change in charge δQ, the corresponding change in occupation number δN is

$$\delta Q = 2e\delta N.$$

Thus equation (8.7) can be rewritten

$$[Q, \Phi] = -\,\mathrm{i}\hbar. \tag{8.8}$$

To derive the analogues of the DC and AC Josephson effects for a flux tunnelling device we shall make use of a very simple description of two coupled superconductors due to Feynman. He suggested taking two pieces of superconductor with order parameters ψ_1 and ψ_2 coupled in some way so as to allow pairs to tunnel between them. In the absence of coupling the phase in each superconductor is arbitrary, and the time-dependent Schrödinger equation

$$\mathrm{i}\hbar\ \mathrm{d}\psi/\mathrm{d}t = E\psi$$

describes the evolution of the order parameter, where E is the energy of either superconductor. Now introduce some coupling k, characteristic of the weak-link region between the two pieces. This will mix in some amplitude of ψ_1 with ψ_2 and vice versa. Thus the Schrödinger equations become two coupled equations

$$\mathrm{i}\hbar\ \mathrm{d}\psi_1/\mathrm{d}t = E_1\psi_1 + k\psi_2 \tag{8.9a}$$

$$\mathrm{i}\hbar\ \mathrm{d}\psi_2/\mathrm{d}t = E_2\psi_2 + k\psi_1. \tag{8.9b}$$

Let us assign states of definite phase, or equivalently flux, to ψ_1 and ψ_2 so that

$$\psi_1 = \psi_{10}\ \exp(\mathrm{i}\theta_1)$$

and

$$\psi_2 = \psi_{20}\ \exp(\mathrm{i}\theta_2)$$

and then substitute these expressions in equations (8.9a) and (8.9b). Equating real and imaginary parts leaves us with four equations which reduce to the following results for the net rate of transfer of electric charge from one superconductor to the other and for the time evolution of the phase difference

$$\theta = \theta_1 - \theta_2$$

of the order parameters in the two pieces:

$$\begin{aligned} \mathrm{d}\psi_1/\mathrm{d}t &= -\,\mathrm{d}\psi_2/\mathrm{d}t = (2k/\hbar)(\psi_{10}\psi_{20})^{1/2}\sin\theta \\ \mathrm{d}\theta_1/\mathrm{d}t &= -\,\mathrm{d}\theta_2/\mathrm{d}t = (k/\hbar)(\psi_{10}/\psi_{20})^{1/2}\cos\theta - qV/2\hbar \end{aligned} \tag{8.10}$$

where V is the direct voltage drop across the weak-link region. In the case of a weak link biased from an external current source $r_1 = r_2 = r$, say, a constant. Thus

$$I = (2k/\hbar)(\psi_{10}\psi_{20})^{1/2}\sin\theta \tag{8.11}$$

which is the DC Josephson relationship, derived in Chapter 2, equation (2.8).

This third derivation of the Josephson effects has been summarised for two reasons: firstly, once one has accepted the essentially quantum mechanical nature of the superconducting state the manipulation required to derive the effects is elementary, requiring only a nodding acquaintance with the time-dependent Schrödinger equation. Secondly, this particular treatment can be readily modified to deal with the other representation of the superconducting state outlined above, which is characterised by a definite pair occupation number N. In this case the order parameter will be written in the form

$$\psi(N) = \langle\theta\rangle\exp(\mathrm{i}N)$$

where $\langle\theta\rangle$ is the average value of the phase of the order parameter throughout the piece of superconductor. The analogue of the weak link between two pieces of superconductor will, in our treatment, be two pieces of superconductor, separated by two weak-link regions, with an inductive ring between them. This arrangement is like the DC SQUID configuration dealt with in Chapter 4. Flux can move through the weak-link regions either into, or out of, the central hole. The flux state of the ring is defined as the difference between the flux applied to the ring from an external source and the actual total flux linking the ring. Of course the coupling of the weak-link regions to flux transfer into or out of the ring increases as the gap between the superconductors gets greater, exactly the opposite of the situation in the conventional electron pair transfer Josephson junction (see figure 8.10).

Using the new definition of the order parameter given above it is simple to relate the flux state of the ring to the difference in the average phase values of the two pieces of superconductor $\theta_1 - \theta_2$:

$$\Phi = \Phi_1 - \Phi_2 = \Phi_0(\theta_1 - \theta_2)/2\pi. \tag{8.12}$$

The weak-link regions serve to hinder the passage of flux from the ring to a lesser or greater extent, depending on their dimensions. This hindering effect can be incorporated into the coupled equations for the order parameters in the two pieces by the inclusion of the parameter k in equations (8.10). Thus if $k=0$ the flux is totally fixed, i.e. the ring is strongly coupled, with no weak-link regions. When k becomes very large flux can move almost unhindered, the situation when the gaps in the ring are very large. By direct comparison with equations (8.10) we have

$$\begin{aligned} \mathrm{d}\psi_1/\mathrm{d}t &= -\,\mathrm{d}\psi_2/\mathrm{d}t = (k/\hbar)(\theta_{10}\theta_{20})^{1/2}\sin[2\pi(N_1-N_2)] \\ \mathrm{d}N_1/\mathrm{d}t &= -\,\mathrm{d}N_2/\mathrm{d}t = (k/\hbar)(\theta_{10}/\theta_{20})^{1/2}\cos[2\pi(N_1-N_2)] - \Phi_0 I/\hbar. \end{aligned} \tag{8.13}$$

As before, the four equations can be solved for the direct voltage V between the two sides of the ring as a function of the occupation number difference $N_1 - N_2$:

$$V = V_1 \sin[2\pi(N_1 - N_2)]. \tag{8.14}$$

When the rate of flux transfer is so high that some critical rate V_1 is exceeded, then a current of pairs will flow from one piece to the other, while flux oscillates in and out of the ring at a frequency f given by the very simple and presumably very accurate relationship

$$\mathrm{d}(N_1 - N_2)/\mathrm{d}t = (2k/\hbar)\sin(\Phi_0 It/\hbar)$$

so that the frequency, f, is

$$f = I/2e \tag{8.15}$$

where I is the direct current flowing and $2e$ is the quantum of charge of a superconducting electron pair.

There has been no indication up to this point about what sort of structure the flux tunnelling weak link should have. No hard and fast answers have yet been given but intuitively one can see that flux tunnelling will be more probable from a ring with a thin wall than from one with a thick wall. It seems that only if the flux distribution within the ring extends outside it with a finite density will there be a finite probability of a tunnelling transition. For this to occur the wall thickness must be of the order of the penetration depth for the superconductor, typically 30 nm for a type I pure metal. The whole ring may be of this thickness or just a small region may be thinned down. The tunnelling probability will similarly depend on the cross-sectional area of the thinned down 'weak'-link region. Just as for Josephson junctions there are two distinct methods for coupling: either by tunnelling of pairs through an insulator or by passage through a metal constriction, so in the flux transfer case the flux quanta may 'tunnel' through a normally forbidden region of superconductor if it is sufficiently thin, or they may pass through a constricted insulating region connecting the interior hole to the exterior. The treatment of flux tunnelling given here is, for reasons of space, necessarily restricted. However the effects may be derived more rigorously using a Lagrangian density treatment (Prance *et al* 1985) which confirms the exact relationships given by equations (8.14) and (8.15).

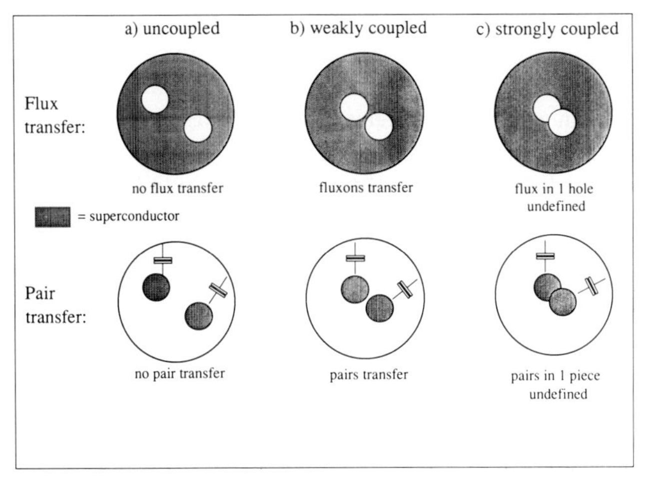

Figure 8.10 Schematic of the conjugate forms for pair and fluxon transfer

If a structure can be produced in which flux tunnelling properties dominate over pair charge tunnelling, what could be done with it? One of the first and most important applications of the Josephson effects was to a measurement of the flux quantum, and then to a new standard of voltage, as we have seen in Chapter 2. Similarly a flux tunnelling junction would allow a direct current to be measured in terms of a frequency and another fundamental constant of physics, e, the electron charge.

8.7 Possible Applications of Flux Tunnelling

In the previous section the essential details of the flux tunnelling model were derived. In this section we shall speculate on some possible applications of these still somewhat hypothetical effects. The relationship between the frequency of flux tunnelling and the DC current flowing through the weak link strongly suggests the possibility of a quantum current standard. This would provide an important complement to the Josephson voltage standard, and the two quantum standards taken together would provide a value for the most important dimensionless fundamental constant of physics, the coupling constant of quantum electrodynamics, the fine structure constant α^{-1}. To realise a quantum current standard it will almost certainly not be sufficient simply to measure the emitted photon frequency from a free running junction voltage-biased into the finite current regime. Although nothing is known about the dissipative processes associated with flux tunnelling it seems likely that the effective resistance of the weak link will be sufficient to make the free running frequency linewidth too great to be useful. Thus some kind of frequency locking to an external source will be required. The possibility of coherent induced flux tunnelling by the application of an external signal has already been considered. The effect of phase locking will be to produce flat step regions in the voltage–current characteristic of the weak link, the steps being equally spaced by current differences of $2eV/\Phi_0$ (see figure 8.11).

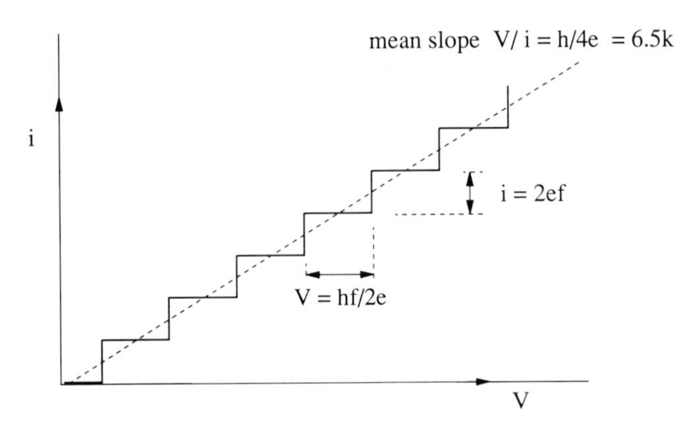

Figure 8.11 Possible I–V characteristic of a microwave-irradiated hybrid weak-link device

A flux tunnelling weak link will have a direct current–voltage characteristic which will be affected by incident electromagnetic radiation and this might form the basis of a useful detector. The characteristic impedance of such a device is not known although it will almost inevitably possess a very low noise temperature. The analogue of a flux detector would be a detector of electric charge, an 'electrometer' in conventional terms. Just as a SQUID is able to detect changes which are only 1 part in 10^5 of a flux quantum, such an electrometer might be able to detect charge changes corresponding to a very small fraction of the charge on a single electron. What is really being detected is a spatial redistribution of charges within one piece of the superconductor.

As for possible logic devices based on single charge transfer, if the weak-link capacitance is, as expected, of order 1 fF the stored energy change associated with a positive logic state would be only

$$\Delta E = (1/2)q^2/C \simeq 2 \times 10^{-23}\ \text{J}$$

compared with an order of 10^{-20} J for a silicon junction.

8.7.1 Single-electron tunnelling devices and Coulomb blockade

A number of groups have attempted to observe the coherent transfer of flux quanta in the conjugate device to a conventional Josephson junction. Clark (1987) has observed the microwave-induced coherent transitions in SQUID ring geometries but other attempts to observe the same effects in singly connected superconducting weak links have so far proved unsuccessful (Ono *et al* 1987). The capacitance of the device must be minimised to observe the desired effects and it is believed that the stray capacitance associated with the bias leads in a singly connected geometry may be sufficient to suppress the quantum coherent effects.

An additional complication to this already complex set of phenomena arose with the proposal and subsequent recent observation of an associated effect. This is known as single-electron tunnelling (SET), first described theoretically by Likharev and Zorin (1985) and which is based on the stochastic tunnelling of single electrons through a tunnel junction, such that the coherence of this transfer process is not involved. The origin of SET arises from the energy band structure described above becoming periodic in e rather than $2e$ when damping of the superconducting weak link is included in the discussion. If the shunt conductance is greater than the quantum value $e^2/\hbar$ then relaxation of the macroscopic states is rapid enough to break the macroscopic aspects of coherence. Stochastic tunnelling of single electrons can be observed particularly if the competing pair-tunnelling effects are suppressed by the application of a large magnetic field. The passage of single tunnelling electrons can be synchronised to an applied microwave field and, for a sufficiently small capacitance, effects analogous to the conjugate Josephson effects should be observable. These have a direct conjugate analogue in the stochastic transfer through a weak link of single flux quanta which was first observed in narrow bridges of granular aluminium (Fiory 1971) and has also been seen in two-dimensional polycrystalline films of high-temperature superconductors (Gallop *et al* 1989).

8.7.2 Turnstile current standard

SET oscillations have now been observed and the basic high-frequency synchronisation effects confirmed. Recently a device has been fabricated which consists of an array of ultrasmall tunnel junctions which may be biased by a symmetrical voltage along the length of the array and simultaneously, via a non-tunnelling capacitative gate, the central electrode, half-way along the array, may also be voltage biased. When an alternating voltage at frequency f is applied to the gate electrode, one electron per cycle of this oscillation is enabled to be transferred through the device. This produces a plateau in the I–V characteristic at $I = ef$ (Geerligs *et al* 1990).

The requirement for accurate transfer of a single electron per cycle of the applied alternating voltage is that the stochastic tunnelling time should be much shorter than a single half-period of the applied oscillation and that the charging energy of the central electrode when it has been charged up by a single electron is greater than a typical thermal fluctuation $kT/2$. Then the central electrode enters a state known as 'Coulomb blockade' in which no further charging is possible and all current flow is suppressed until the single charging electron is able to transfer through the device when the applied voltage is reduced. Geerligs and co-workers have shown that, for an applied oscillation in the radio-frequency region, induced constant-current steps with the expected spacing of fe are observable with an accuracy of 0.1%. The likelihood that a quantum current standard with sufficient accuracy (say < 1 in 10^7) to be useful can be made using SET rather than Bloch oscillations seems low at present. However, if tunnel junctions with capacitance $C < 1$ aF could be routinely made then the position would change significantly. Effects of this type were first observed by Giaver and Zeller (1968) in a system of superconducting particles embedded in an insulating matrix.

8.7.3 Possible corrections to the exact frequency and voltage or current relationships

Section 2.4.1 of Chapter 2 dealt with possible corrections to the basic 'classical' Josephson voltage–frequency relationship. No corrections are expected above the level of about 1 part in 10^{15}, provided that the geometry and dimensions of the junction and its associated leads are suitably chosen. In the light of the more realistic quantum conjugate variable treatment given in the preceding section it is necessary to re-examine this question. At the most fundamental level the voltage–frequency and current–frequency relationships relate the periodicity in time of the order parameter to the applied voltage or current. The time variation of the order parameter is phase-locked to a strong external electromagnetic signal. If, in the case of the conventional Josephson effect, the weak-link junction allowed every so often a flux quantum to tunnel through it orthogonally to the direction of current flow, the order parameter will undergo a change of 2π which will be unrelated to the value of direct voltage applied. So if the rate of flux tunnelling was r (s^{-1}) and the applied frequency was f (Hz) the fractional error produced would be r/f. Of course all the flux quanta would have to tunnel in the same sense through the weak link to produce an error as great as this. In the event of the tunnelling being a random process, significant averaging would occur. Conversely in the case of a current standard (which as we have discussed would only be feasible if a weak link could be produced in which flux tunnelling rather than pair tunnelling was the dominant mode), any pair tunnelling present will introduce small corrections to the current–frequency relationship (equation (8.15)).

9

High-temperature Superconductors and SQUIDs

9.1 The Superconductor Revolution

While this book was being written the tempo of research into SQUIDs, and superconducting electronics in general, was raised dramatically. It has been estimated that the numbers of people involved in full-time research into superconductivity in general increased by an order of magnitude between 1987 and 1989. This arose from the discovery by Bednorz and Müller (1986) of a completely new class of superconductors made up of ternary and quaternary copper oxides. (The discoverers were awarded the Nobel prize the following year. This unprecedented turn of speed by the award committee is regarded by many as recognition of the expected dramatic technological impact of their work.)

9.1.1 The perovskite superconductors

These ceramic materials seemed very improbable superconductors at the time of their discovery, exhibiting many other contradictory properties, such as antiferromagnetism and semiconducting or insulating properties when their compositions, dopant levels or oxygen content are only very slightly changed. However, when made as a single phase with the correct composition and structure genuine superconductivity with both true zero resistance and a surprisingly complete Meissner effect are observed in these high-temperature superconductors (HTSs). The generic form of the first-generation oxide superconductors may be written as RXCuO where R is almost any rare earth metal and X is typically an element from group 2A. The first compound found (R = La, X = Sr), also known by the abbreviation LSCO, had a single layer of ordered CuO planes separated by layers containing La and Ba oxide layers with a repeat distance in the *c* axis perpendicular to the planes of some 1.32 nm. The maximum T_c found was around 50 K.

A second compound with $T_c > 91$ K, 14 K above the normal boiling point of liquid nitrogen, had the composition $YBa_2Cu_3O_7$. Again the oxygen is present in ordered form, coordinated in a planar structure with the copper atoms, linear chains then connecting to the R and X sites. A diagram of the structure of the most used compound to date, $YBa_2Cu_3O_7$ (known universally by the abbreviation YBCO), is shown in figure 9.1(*a*). The structure is similar to that of perovskite, although the ternary nature and the ordering ensure that the unit cell is orthorhombic and some three times longer in the *c* axis direction than in the *a* and *b* axes. Some 18 months after Bednorz and Müller's original discovery, two new compounds were discovered which further extended upwards the transition temperatures of these copper oxide ceramic materials. These did not contain any rare earth element. At the time of writing the compounds $TlCaBaCuO_x$ and $BiSrCaCuO_x$ have pushed up the (confirmed) T_c levels to 125 K and 110 K respectively for the value at which true zero resistance can be attained (see figure 9.1(*b*)). Some properties of a number of the new perovskite superconductors are summarised in table 9.1.

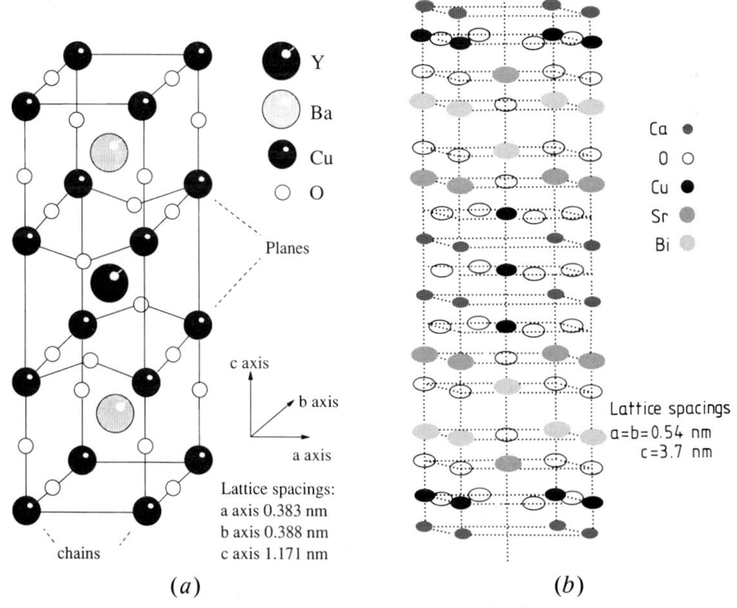

Figure 9.1 Structure of (*a*) $YBa_2Cu_3O_7$, and (*b*) $Bi_2Sr_2Ca_2Cu_2O_x$

Table 9.1 Properties of the new perovskite superconductors

Composition	Transition temperature, T_c (K)	Penetration depth, λ (nm)	Coherence length, ξ (nm)	Upper critical field, B_{c2} (T)
$(LaSr)_2CuO_4$	38	—	3.7 (*ab*)	125(*ab*)
			0.7 (*c*)	24(*c*)
$YBa_2Cu_3O_7$	93	140	3.1 (*ab*)	210(*ab*)
			0.51(*c*)	35(*c*)
$Bi_2Sr_2Ca_2Cu_3O_{10}$	110	180	4 (*ab*)	400(*ab*)
			0.21(*c*)	19(*c*)
$Tl_2Ba_2Ca_2Cu_3O_{10}$	125	210	—	—

9.2 Advantages of High-temperature Superconductivity for the Josephson Effects and SQUIDs

Immediately following the discovery by Bednorz and Müller, work began in earnest to use the new materials to fabricate SQUIDs. The expected advantages have been clear from the start. A higher transition temperature brings simplification and cost saving in refrigeration, just as it should for large-scale electrical engineering applications such as power transmission. However in the small-scale arena several other advantages exist in addition.

We saw in Chapter 6 that high-frequency applications are limited by an upper frequency which scales with the superconducting energy gap. Although it appears that the microscopic BCS theory, which so effectively explained almost all pre-1986 superconductors, may not be entirely adequate to describe these ceramics, it nevertheless seems to be the case that whatever theory replaces BCS will also require an energy gap that scales approximately with T_c. If this is so the resulting ten-fold increase in this parameter may be expected to yield detectors and oscillators whose frequency ranges extend into the near-infrared, perhaps to wavelengths as short as 5–10 μm. (Such performance has yet to be realised.) Similarly, even higher switching speeds for logic gates would be expected and extensive new efforts in digital electronics are underway again in the Josephson computer and signal processing fields.

9.2.1 Effects of higher operating temperature on performance

The possibility of using SQUIDs and other Josephson devices at much higher operating temperatures (perhaps even at room temperature in the future, if more dramatic improvements in novel materials can be found) is the driving advantage for the development of HTS devices. It is generally assumed that there is an inevitable disadvantage to this higher-temperature operation. This arises from an increase in Nyquist noise associated with the higher temperature of dissipative elements in electronic circuits within the superconducting devices, or magnetically coupled to them. For example, for a Josephson oscillator the linewidth of emitted electromagnetic radiation is proportional to operating temperature T:

$$\mathrm{d}f = 4kTR/\Phi_0^2 \tag{2.24}$$

(see section 2.3). For uses such as low-power local oscillators for HTS Josephson heterodyne detectors the *fractional* linewidth at the highest operating frequencies might be expected to be comparable to that of conventional superconducting devices, since the upper frequency limit of Josephson oscillators also scales linearly with T_c, and hence with operating temperature.

Similarly with SQUIDs of both RF and DC varieties (see sections 3.3 and 4.3, respectively) the classical Nyquist noise treatment suggests that the voltage and current noise spectral densities are proportional to T. Thus we may expect non-ideal SQUIDs, for which a classical treatment is adequate, to exhibit increased noise due to higher operating temperature. However, the quantum mechanical treatment of DC SQUIDs shows, and this is borne out by experimental results on very small-area low-inductance devices, that ideally the noise becomes temperature-independent. To achieve this happy situation one requires a SQUID for which $\hbar\omega > kT$. Since $\omega \sim 1/(LC)^{1/2}$, extraordinarily small capacitance junctions will be required to realise this limit, given that T is at least ten times greater with the new materials than for conventional superconductors. Assuming L cannot be practically reduced, while maintaining adequately close coupling to a signal coil, C will require to be reduced to around 10^{-16} F to attain this limit. As we shall see below, no Josephson tunnel junction has yet been fabricated using the new high-temperature materials so the problem of producing such a small and well characterised structure will clearly be extraordinarily challenging. However it would be premature to insist that this will never be achieved, and hence the increased Nyquist noise associated with higher operating temperatures does not seem to represent a fundamental limitation. In principle, quantum-limited SQUIDs could operate at any temperature.

9.2.2 The fundamental limit to HTS SQUID sensitivity when operated at 77 K

The assumption seems to be widespread that HTS SQUIDs operated at, say, 77 K will inevitably be less sensitive than their conventional superconductor equivalents operated at 4.2 K. With this point in mind it is interesting to consider various design criteria for DC SQUIDs in the light of the properties of the Josephson junctions employed. Let us assume that the naive resistively shunted junction (RSJ) model provides an adequate description of a DC SQUID, with the usual device parameters. First, to attain a large flux to voltage transfer function we require:

$$\beta = 2\pi L i_1/\Phi_0 \sim 1 \tag{9.1}$$

where L is the loop inductance and i_1 is the critical current of each of the two weak links (assumed to be identical for simplicity). It is also important that the I–V characteristic should exhibit no hysteresis which is achieved by providing adequate damping, in the form of a shunt resistance, R, and capacitance, C, satisfying the condition

$$\beta_c = 2\pi i_1 R^2 C/\Phi_0 < 1. \tag{9.2}$$

The Josephson coupling energy of each weak link must be greater than that of a typical thermal fluctuation, otherwise there will be no coherence. The dimensionless parameter Γ is constrained thus:

$$\Gamma = 2\pi kT/\Phi_0 i_1 \ll 1. \tag{9.3}$$

The magnetic energy difference between adjacent fluxon winding numbers of the SQUID ring must also be greater than that of a typical thermal fluctuation, otherwise the flux number will be undefined:

$$\Phi_0{}^2/2L \gg kT. \tag{9.4}$$

We assume the thermal noise spectral density is white, but to incorporate quantum mechanics in this analysis, in a somewaht naive way, we will include a term representing the zero-point energy, so the resulting total noise spectral density is

$$S_E = (kT\Phi_0/i_1 R + \hbar/2). \tag{9.5}$$

A quantum limited HTS SQUID operating at 77 K is possible in principle if the second term in equation (9.5) dominates the first, requiring that

$$e i_1 R/kT > 1. \tag{9.6}$$

According to the Josephson tunnelling calculations based on conventional BCS theory

$$i_1 R \sim 2\Delta/e$$

so condition (9.6) becomes

$$2\Delta/kT > 1$$

which is easily satisfied if near-perfect tunnel junctions can be made.

The other conditions still remain to be satisfied. Equation (9.4) requires that for operation at 77 K

$$L \ll \Phi_0{}^2/2kT \qquad \text{or} \qquad L < 1\ \text{nH}$$

whereas the preceding condition demands that

$$i_1 \gg 2\pi k_B T/\Phi_0 \qquad \text{or} \qquad i_1 > 3\ \mu\text{A}$$

for operation at the same temperature. Evaluating equation (9.6) with this estimate leads to

$$R > kT/2ei_1 \qquad \text{or} \qquad R > 1000\ \Omega.$$

Inserting the numerical value for Φ_0 in expression (9.1) leads to

$$Li_1 \sim 1 \times 10^{-15}\ \text{Wb}$$

and this is compatible with the previous conditions. Finally, and perhaps most difficult to achieve, condition (9.2) requires

$$C < \Phi_0/2\pi i_1 R^2 \qquad \text{or} \qquad C < 10^{-16}\ \text{F}.$$

This figure is several hundred times smaller than an upper limit estimate of the capacitance of natural intergrain junctions (Gallop *et al* 1989). While it is hard to imagine attaining such a value, given the still primitive HTS junction fabrication technology, it is only an order of magnitude smaller than the value already achieved with conventional superconducting tunnel junctions (Fulton *et al* 1987) and should not be regarded as permanently impossible. The conclusion from this brief discussion is that, while HTS SQUIDs operating at 77 K are likely to remain noisier than other SQUIDs operating at lower temperatures, this is not an inevitable situation if the fabrication of HTS Josephson junctions can be perfected.

9.3 SQUIDs made from Sintered Ceramic

The YBCO compound, the most widely investigated high-temperature superconductor, possesses a T_c of 93 K, an almost exact ten-fold increase compared with Nb, the workhorse material previously used for SQUIDs. The first reported SQUIDs were made from bulk pieces of the ceramic yttrium compound, using weak links which already existed between grains of the polycrystalline sintered material (Colclough *et al* 1987). The observed low critical current densities of sintered samples have been explained by the early realisation that these specimens consist of an agglomeration of small ($\leqslant 10\ \mu$m sized) grains which are only weakly coupled together so that each grain boundary acts as a weak link. Although this is a major problem for conventional uses requiring high critical current densities, it proved to be a lucky chance which allowed HTS SQUIDs to be very rapidly constructed.

The first step was taken by Gough *et al* (1987), who demonstrated the existence of flux quantisation in a ring of the sintered superconductor. The low critical current density of the material meant that it was relatively easy to induce transitions in the quantised flux state of the ring, using electromagnetic pulses. The amplitude of each flux change was measured by a conventional SQUID magnetometer coupled to the ring but maintained at liquid helium temperatures, a calibration being provided by a long thin toroidal coil which linked the ring. The distribution in magnitude of the flux changes produced was examined and it was found that all were a multiple of the smallest observed change $\delta\Phi = 0.95(\pm 0.1)h/2e$. This provides convincing evidence that paired elementary charges ($|2e|$) are responsible for superconductivity in these ceramics, just as in normal superconductors.

The work of the same group showed, unexpectedly, that single pieces of superconductor showed SQUID operation with only a few discrete magnetic field periodicities, without any ring geometry being defined. It is assumed that just a few dominant naturally occurring loops, incorporating many inter-grain junctions, exist within the material, although these effects are thought by the author to be confined to a weakly superconducting surface perimeter region (Gallop *et al* 1988).

Other RF and DC SQUID designs rapidly followed, in which again the weak links were to be found between grains (typically a few microns in linear dimensions). It was only necessary to cut away the bulk material to 'neck-down' a region to a typical size of around 0.1 mm to ensure that only a few weak links were present. This was fortunate since the coherence length in the ceramic superconductors is extremely short (and anisotropic, <3 nm in the a and b axis directions of the crystal, but <0.3 nm in the c direction), and even the most sophisticated lithography presently available is not capable of producing microbridges of this size in favourable materials, let alone these complex ceramics. The bulk SQUIDs operating at 77 K demonstrated surprisingly good noise performance, comparable with that of commercial SQUID systems operating at 4.2 K, with $\delta\Phi \sim 1 \times 10^{-4}\Phi_0\ \mathrm{Hz}^{-1/2}$ sensitivity. Fortuitously these SQUIDs are also remarkably robust.

Another design has arisen from Zimmerman *et al*'s (1987) early work which a number of groups have repeated, in which a single-junction RF SQUID is formed using a weak-link 'break junction'. A disc of the sintered superconductor is drilled with a hole (~ 1 mm) and a slot is cut radially from the perimeter towards the hole, leaving a small bridge of material. The resulting ring is then fractured at the slot by inserting a tapered pin into it at low temperatures. Then a 'point-contact' Josephson junction is formed by slightly withdrawing the pin, allowing the clean but irregular surfaces thus formed on either side of the break to touch lightly (see figure 9.2).

The main reason why these devices made from bulk sintered material are not already more widely used is that to date the necessary input coils and flux transformers have not been developed. This is due to the lack of availability of flexible HTS wire and additional difficulties of fabricating thin or thick film planar coils to perform the same function. This is an important area where further work is urgently needed (see section 9.3.2).

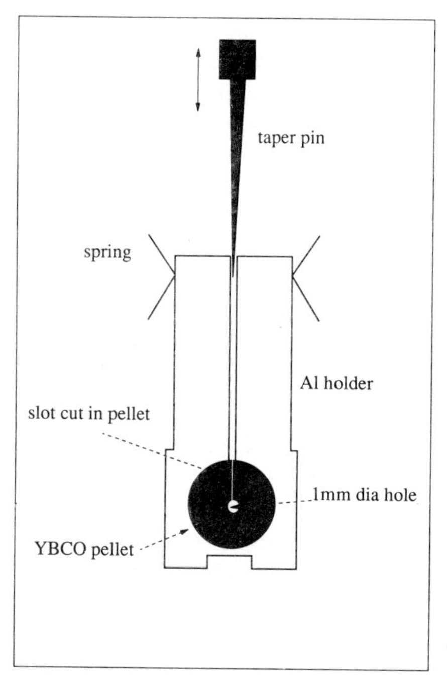

Figure 9.2 YBCO 'break junction' RF SQUID (after Zimmerman *et al* 1987)

9.3.1 Thin film SQUIDs using high-temperature superconductors

The observation that bulk sintered ceramic samples only had a critical current density of $\sim 10^3\ \mathrm{A\,cm^{-2}}$ at 77 K proved a serious limitation on any proposed high-power application. This prompted rapid development of thin film deposition techniques which succeeded in achieving a thousand-fold increase in j_c by epitaxial deposition on $SrTiO_3$ substrates. Other cheaper monocrystalline substrates with better electrical properties such as MgO and zirconia have also been widely used to produce epitaxial, polycrystalline or amorphous films. The same technologies have been employed to produce thin film SQUIDs.

The short coherence length, the complex chemistry of the compounds and the need to oxygenate the compounds at temperatures of order 600 °C after deposition have prevented the production of artificial Josephson tunnelling barriers, and hence designed Josephson junctions of this type (but see the following section for details of a fabricated quasi-particle junction). As the emphasis on HTS material fabrication development has moved towards thin film SQUIDs, natural weak links in polycrystalline films are becoming preferred already to bulk structure devices. Using conventional photolithography and either wet or dry etching, it has proved relatively straightforward to pattern small-scale features suitable for SQUID loops and weak links containing only a few grain-boundary junctions. Patterning by laser ablation has also been widely used, although generally there seems to be some damage to the film edges when this method is employed. The work of Foglietti *et al* (1989) at IBM has probably been most successful to date in this area of low-noise HTS thin film devices. At the time of writing they have succeeded in producing a TlBaCaCuO DC SQUID with an energy sensitivity of 6×10^{-30} J Hz^{-1} at 77 K, substantially better than commercially available conventional superconductor SQUIDs operated at 4.2 K. So far $1/f$ noise does not seem to be a particular problem with this material. The performance of this Tl SQUID must be seen as one of the key results on which is based the widely held, perhaps optimistic view, that HTS SQUID applications are close to realisation, at least in prototype form. The same group has also recently reported fabricating the first simple three-layer Josephson junction, consisting of two HTS thin films separated by a uniform insulating layer only 1 nm thick (Laibowitz *et al* 1990). These junctions do not exhibit ideal or even reproducible properties at present, with no energy gap structure visible. Much improved junction properties will be required for high-frequency or fast-logic applications. However, for situations where only one or two Josephson junctions are required it has proved possible to produce DC SQUIDs using fabricated 'edge junctions' in which a deposited thin film of YBaCuO is covered with a protective layer of BaF. A raked edge is produced on this two-layer film by ion-beam etching before it is exposed to an oxy-flouridising plasma which provides the thin insulating layer. Finally an overlay of YBCO is deposited over the edge and is patterned to give two protruding fingers (about 5 μm wide) which complete a superconducting ring (20 μm × 20 μm). This achievement of fabricated three-level Josephson junctions must be seen as a considerable breakthrough although to date it has proved more practical to make use of naturally occuring weak links.

Another promising route to junctions has been devised at IBM, in which a bi-crystal substrate is made by joining two pieces of single-crystal substrate, e.g. MgO, with a well-defined angle of misalignment between their crystal axes. An epitaxial film is then deposited over the join and the width of the film across this artificial grain boundary can be reduced to an appropriate value to provide the required critical current (j_c is a rather well-defined function of the misalignment angle). Gross *et al* (1990) claim that this is an even more reliable method than the edge junction technique and they have produced a YBCO DC SQUID which again has a noise performance which is as good at 77 K as that of most commercially available Nb SQUIDs at 4.2 K (see figure 9.3).

Other techniques which have also produced reasonable junctions involve introducing a step of height comparable with the film thickness in the substrate (~300 nm) and then patterning the film where it has been deposited over the step region. It should be stressed that all of the junctions produced by these methods are still far from ideal, with I_cR products only of the order of 1 mV or less at 77 K, rather than the hoped for 10–20 mV. However they seem stable and very suitable at least for prototype SQUIDs.

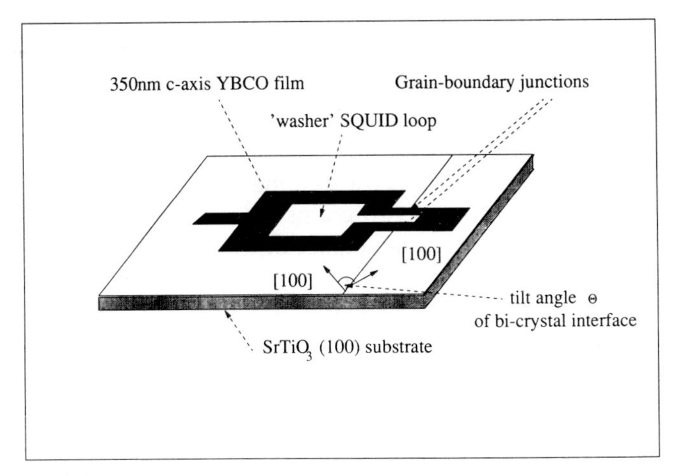

Figure 9.3 Grain-boundary DC SQUID with junctions fabricated at a bi-crystal interface

9.3.2 Planar coils and flux noise

There is a price to be paid for the convenience of the natural Josephson junctions existing at each grain boundary in polycrystalline thin HTS films. It was noticed early on in the development of our topic that the cuprate superconductors are of extreme type II. That is to say it is energetically favourable for flux lines to penetrate the material when the lower critical field H_{c1} is exceeded. For the new materials $H_{c1} \sim 1$ mT is considerably lower than for most conventional superconductors. (Fortunately H_{c2} is actually considerably higher than for the Nb-based alloy superconductors, since high-current, field or power applications will require this.) The effective H_{c1} for polycrystalline specimens is even lower since the weakly superconducting inter-grain boundaries form a network of sites where flux lines may readily penetrate the material.

Even though it is straightforward to maintain the ambient magnetic field well below the H_{c1} value for polycrystalline specimens this is not a sufficient precaution to ensure that flux lines will not be trapped through thin films as they are cooled through T_c. The cooling process, particularly if it is rapid or inhomogeneous, is liable to produce circulating supercurrents, driven possibly by minute thermoelectric EMFs which are capable of establishing significant circulating currents in the very low-resistance proto-superconducting state which exists within the width of the superconducting transition. Furthermore currents in the plane of the thin film generate rather high magnetic fields at the edge, due to the sharp curvature of the film here, which may locally exceed H_{c1}. For these and other reasons thin polycrystalline films (at least of conventional superconducting materials) invariably are penetrated by an irregular pattern of trapped flux lines.

Assuming that the irregular trapped flux line pattern also exists in the new high-temperature superconductors there are significant implications. The depth of a potential energy well which defines the metastable local minimum in which each fluxon is trapped may be very low. (If the critical current of a weak link between two grains is i_1 the well depth will be only of order $Li_1{}^2/2$, where L is the inductance of the non-superconducting hole in which the flux line is trapped.) i_1 will of course be temperature-dependent and a particular grain boundary will only be effective in trapping a flux line if $Li_1{}^2/2 \gg kT$, so that thermal fluctuations are not capable of rapidly exciting flux lines over the barrier so that they can move from one local minimum to an adjoining lower-energy one. Since it is assumed that there is a wide range of i_1 values present in any polycrystalline material at any temperature it seems likely that there will always be some weak links where flux pinning energy is comparable with kT and stochastic motion of flux lines must be expected. This is equivalent to 'flux creep' observed in conventional superconductors at higher applied fields.

The importance of this discussion for SQUID systems is probably obvious by now. A thin film flux transformer or other input coil configuration coupled to a SQUID ring will almost certainly be penetrated by flux lines which at the operating temperature will not all be securely pinned. The resulting flux jumps will generate spurious noise at the SQUID itself. Experimental observation of this noise in YBCO thin films has been given by at least two groups (Ferrari *et al* 1989, Irie *et al* 1989; see figure 9.4) and they both observe excessively large noise with a disturbing approximately $1/f$ spectral distribution.

9.3.3 Epitaxial films, multilayers and flux transformers

In spite of the relatively low critical current densities achieved with bulk high-temperature superconductors, thin films of the same materials can now be made which possess very respectable j_c values; as high as 10^7 A cm^{-2} even at 77 K. This performance has been achieved through the production of essentially single-crystal thin films grown epitaxially on smooth single-crystal substrates of a variety of materials (usually with the c axis perpendicular to the substrate). Deposition by laser ablation, sputtering or co-evaporation have all been used successfully, the essential feature in common to these '*in situ*' techniques being that the material is deposited onto a hot (~700 °C) substrate in the presence of active oxygen, so that the thin film superconducts directly on cooling with no further annealing or oxygenation process being required.

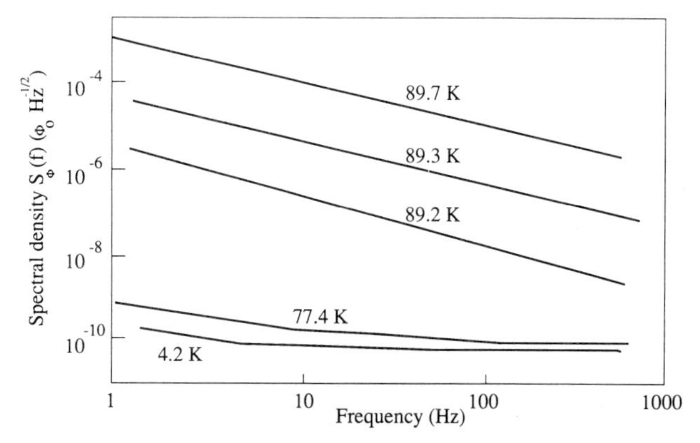

Figure 9.4 Spectral density of flux noise for an epitaxial YBCO thin film at five different temperatures (after Ferrari *et al* 1989)

Progress with multilayers in which epitaxy can be maintained from the substrate throughout successive film depositions has been reported by several groups. Geerk *et al* (1989) have reviewed the present state of tunnelling junction data. There is a fairly univeral observation that the presence of a gap (~20 mV) is clear, but that even in the highest quality junctions it accounts for only a 20% change in dI/dV. (In tunnel junctions with Pb counter electrodes the leakage below the Pb gap at low temperature is very low, implying that the tunnel barrier is continuous.) Also the T-dependence of the gap, when corrected for temperature smearing of the Fermi levels, is almost undetectable except very close to T_c, and certainly is not BCS-like. This applies to both c- and a-axis oriented films. Also doping or reducing the oxygen content to reduce T_c only serves to reduce the amplitude of gap features but does not change their energy.

Another major breakthrough has come with the development of trilayer circuits which form planar flux transformers. As so often before in the SQUID field, Clarke's group at Berkeley seem to be in the lead in this area (Wellstood *et al* 1990). They have used $SrTiO_3$ as an insulator which may be deposited epitaxially on laser ablated YBCO, to give excellent insulation resistance at low temperatures, forming the 'cross-under' for the flux transformer (see figure 9.5). On top of the insulator a second YBCO layer is deposited, which though not having as high a j_c value as the lower film is nevertheless good enough to have low-noise properties (i.e. the flux lines are rather well pinned). The contact area between the two YBCO films provides a superconducting interface which can support reasonable persistent currents (~ mA). This is a most important step towards real HTS SQUID applications.

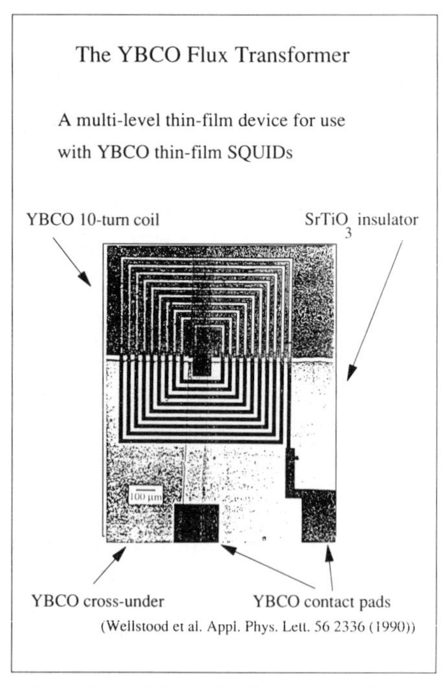

Figure 9.5 The YBCO flux transformer

9.4 Josephson Devices made with High-temperature Superconductors

In addition to work on SQUIDs, results of Josephson behaviour in singly connected HTS junctions have also been given. Again almost all work has relied on natural weak links between grains in polycrystalline or granular specimens. The AC Josephson effects are present in HTS materials, as has been demonstrated by microwave-induced constant-voltage steps (see section 2.4) in the current–voltage characteristic of a 'necked-down' region of sintered YBCO or LSCO.

Witt (1988) has made measurements of the basic voltage step spacing when the weak link(s) is/are irradiated at frequency f and has shown that the value of $h/2e$ agrees with that in conventional materials to within 1 part in 10^5. (This is further confirmation that paired charge carriers are responsible for superconductivity in the cuprate superconductors.)

There can be little doubt that the very serious difficulties attached to fabricating 'designer' Josephson tunnel junctions from the new materials will be solved within the next few years. There have already been suggestions, and preliminary investigations have begun, that the proximity effect may be used to reduce the difficulty presented by the extremely short coherence lengths which the perovskite superconductors exhibit. The proximity effect characterises the behaviour of an interface region between a superconductor and a non-superconductor. The decay of the superconducting order parameter is not instantaneous at the interface but there is a region near it in the superconductor in which it is depressed (the superconductivity being thereby weakened), whereas in the non-superconductor, which could be a semiconductor or even an insulator as well as the more usual case of a normal metal, the extension of a finite order parameter into it renders a region of the material a superconductor.

9.4.1 Fabricated tunnel barrier

At the time of writing only one group has reported a tunnel junction in which both electrodes are made of high-temperature superconductor.

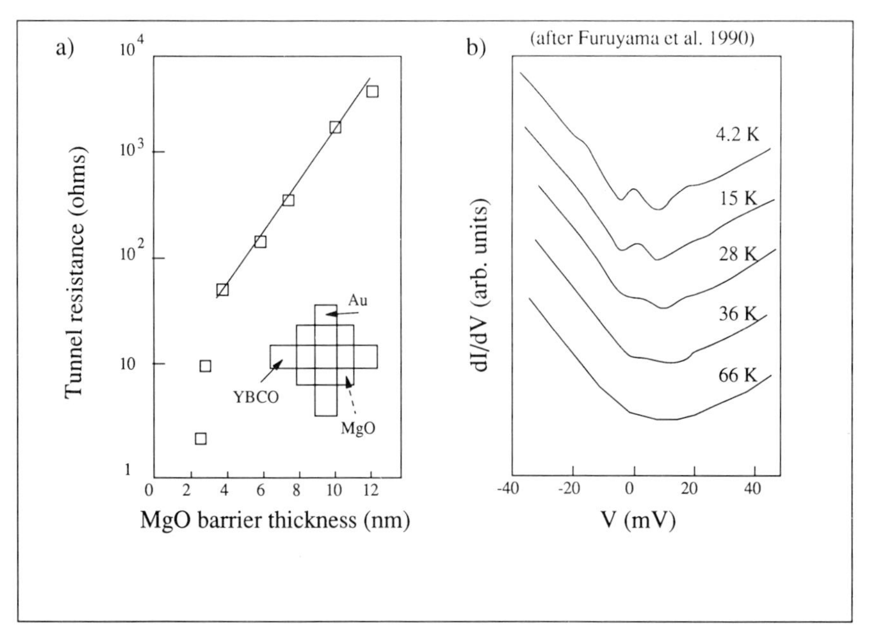

Figure 9.6 YBCO/MgO/Au tunnel junction: (*a*) resistance versus barrier thickness; and (*b*) conductance versus voltage at various temperatures

Shiota *et al* (1989) have produced a planar YBCO–YBCO sandwich whose I–V characteristic shows a very clear superconducting gap structure ($\Delta(T)/e = 18$ meV at 4.2 K), but whose inter-electrode coupling is too weak to show a Josephson supercurrent at zero bias. A YBCO film was first RF magnetron sputtered onto an MgO substrate. The novelty of this approach came with the use of CF_4 gas plasma 'fluoridation' of the surface of this film to render it insulating. Finally an upper YBCO electrode was again deposited by sputtering, each film being about 1 μm thick. The authors suggest that the key to their success with this method is that the insulating barrier is produced in the surface of the superconductor itself, so that deposition of a second layer occurs essentially onto a substrate of almost identical crystal structure. Measurement of the temperature dependence of the gap parameter showed that it reached zero at $T \sim 77$ K.

9.5 Conclusions

The final two chapters in this book were intended to emphasise the essentially open-ended nature of the present state of superconducting electronics. The existence of superconductivity has been known for more than eighty years, yet the subject regularly brings about dramatic surprises every decade or so. An understanding at the microscopic level took almost fifty years and the impact on science and technology of weak superconductivity in helium-cooled materials is still in progress. Now the whole subject has been stirred up again with the discovery of technologically important superconductors at much higher temperatures. Only an extreme pessimist would expect our present lack of knowledge of the microscopic origins of this new phenomenon to take another fifty years to unravel. However, even if this were the case, we may be confident that superconducting devices and electronics will continue to surprise us with many novel and challenging applications.

Appendix A

Electrodynamics of Superconductors

Accepting that a superconductor can be described by a complex order parameter, a kind of macroscopic wave function, we discuss in the first appendix how this order parameter responds to applied electric and magnetic fields, by analogy with the influence of such fields on single-particle wave functions. This is obviously not a rigorous derivation and its validity, like every other aspect of the subject, depends ultimately on the experimental verification of the model's predictions.

A.1 Classical Electrodynamics of a Charged Particle

An electrically uncharged particle possessing no higher-order multipole electric or magnetic charge distributions is quite unaffected by weak applied fields. In this case the definition 'weak' implies that the field strength is not sufficient to affect the wave function amplitude significantly. Suppose now that this point particle has a charge q. The Lorentz force law tells us how a non-relativistic particle of mass m moves if it has velocity $\boldsymbol{v}$, in applied electric and magnetic fields $\boldsymbol{E}$ and $\boldsymbol{B}$. The force $\boldsymbol{F}$ on it is given by

$$\boldsymbol{F} = m\,\mathrm{d}\boldsymbol{v}/\mathrm{d}t = q(\boldsymbol{E} + \boldsymbol{v} \times \boldsymbol{B}). \tag{A.1}$$

Classically, the assumption is that $\boldsymbol{E}$ and $\boldsymbol{B}$ represent the real physical observables of the electromagnetic field. Both $\boldsymbol{E}$ and $\boldsymbol{B}$ can be derived from potential functions, but these potentials are merely aids to calculation. The two relevant Maxwell equations are

$$\nabla \times \boldsymbol{E} + \partial \boldsymbol{B}/\partial t = 0 \tag{A.2}$$

$$\nabla \cdot \boldsymbol{B} - 0. \tag{A.3}$$

The vector identities $\nabla \times (\nabla\chi) = 0$ for any scalar χ, and $\nabla \cdot (\nabla \times \boldsymbol{U}) = 0$ for any vector $\boldsymbol{U}$ prove useful. The first shows that, since $\nabla \times \boldsymbol{E} \neq 0$, $\boldsymbol{E}$ cannot be expressed as only the gradient of a scalar. On the other hand the second identity shows that, since $\nabla \cdot \boldsymbol{B} = 0$, $\boldsymbol{B}$ can be represented by the curl of a vector potential $\boldsymbol{A}$. Substituting for $\boldsymbol{B}$ in equation (A.2) gives

$$\nabla \times (\boldsymbol{E} + \partial \boldsymbol{A}/\partial t) = 0. \tag{A.4}$$

Introducing a scalar potential function V defined by

$$\boldsymbol{E} = -\nabla V - \partial \boldsymbol{A}/\partial t \tag{A.5}$$

the Lorentz force law may be rewritten only in terms of the scalar and vector potentials:

$$\boldsymbol{F} = q[-\nabla V - \partial \boldsymbol{A}/\partial t + (\boldsymbol{v} \times \nabla \times \boldsymbol{A})]. \tag{A.6}$$

A.2 The Generalised Momentum and Schrödinger's Equation in a Magnetic Field

The momentum of a single charged particle in a magnetic field is not simply $m\boldsymbol{v}$, as it would be in the absence of any magnetic vector potential $\boldsymbol{A}$. It is convenient to define the Lagrangian function L for the system as the difference of the kinetic energy T and the potential energy U. The total 'generalised momentum' can be obtained from L for the system:

$$L = mv^2/2 - qV + q\boldsymbol{A} \cdot \boldsymbol{v}. \tag{A.7}$$

The first term, representing the particle's kinetic energy, is straightforward. The second term is the electrostatic potential energy whereas the third term is the potential energy arising from the motion of the particle in a magnetic field. Its form may be derived from the requirement that L be Lorentz invariant. The relationship between generalised momentum $\boldsymbol{P}$ and L is well known from classical mechanics (see for example Goldstein 1969)

$$\boldsymbol{P} = \partial L/\partial \boldsymbol{v} = m\boldsymbol{v} + q\boldsymbol{A}. \tag{A.8}$$

Thus an additional contribution $q\boldsymbol{A}$ arises from the momentum associated with the magnetic field itself. Classically, momentum can be divided into two contributions: kinetic momentum which corresponds to mass times velocity, and a contribution $q\boldsymbol{A}$ which arises from the effect of the vector potential on the charged particle. For a particle in a uniform magnetic field the vector potential does not depend on any coordinates which are orthogonal to the field. This requires that each generalised momentum component in this plane must be conserved. Since each component of kinetic momentum in this plane is clearly not conserved when the particle is moving in a circle, a contribution from the vector potential is required, of magnitude qA.

In making the transition from classical to quantum mechanics, canonical coordinates and momenta are replaced by the corresponding operators. The classical Hamiltonian for a charged particle is

$$\hat{H} = p^2/2m + qV. \tag{A.9}$$

The operator corresponding to kinetic momentum is $-i\hbar\nabla$ so that the quantum mechanical Hamiltonian becomes

$$\hat{H} = -(\hbar^2/2m)(\nabla - qA/\hbar)^2 + qV. \tag{A.10}$$

The time-dependent Schrödinger equation for this charged particle in a magnetic field is

$$-i\hbar\partial\psi/\partial t = [-(\hbar^2/2m)(\nabla - qA/\hbar)^2 + q\varphi]\psi. \tag{A.11}$$

For a stationary state ψ_n with energy eigenvalue E_n we know that

$$\hat{H}\psi_n = E_n\psi_n.$$

There is another way of calculating the effect of a magnetic field on the wave function of a single charged particle, using the so-called 'path integral' formulation of quantum mechanics. This is directly related to Hamilton's principle of least action, in classical mechanics. The path followed by a dynamical system is such that the action integral $S = \int L(s,t)$ is a minimum.

For a quantum mechanical system the path is not defined but it has been shown that the action integral represents the phase change of the wave function. The contribution to S arising from the magnetic field was shown above to be $q\boldsymbol{A}\cdot\boldsymbol{v}$ so that the vector potential contribution to the wave function is to multiply it by a term

$$\exp\left(iq/\hbar \int \boldsymbol{A}\cdot \mathrm{d}\boldsymbol{s}\right). \tag{A.12}$$

This multiplying factor has modulus one confirming that the term corresponds merely to a phase shift of ψ. It is relatively simple to show that the path integral form is exactly equivalent to the derivation based on a quantum mechanical modification of the classical Lagrangian for a charged particle in a combination of electric and magnetic fields.

A fundamental axiom of quantum mechanics is that wave functions should be single valued, and this same axiom can be applied to the macroscopic wave function describing a superconductor. Thus, making use of the path integral formalism, the effect of an applied magnetic field on a superconductor is to produce the phase shift described by equation (A.12). Around any closed path within the body of the superconductor the total phase change must be $2n\pi$, where n is an integer. This is not the end of the argument, though. Imagine that the contour is reduced smoothly towards zero length. The integer n cannot jump suddenly from one value to another from one contour to an infinitesimally different one. For the truly zero-length contour the phase change must be zero, and the value of n zero. Thus by continuity, the phase change on any closed contour which remains within the body of the superconductor and can be reduced smoothly to a zero-length contour (i.e. the superconductor must be simply connected, so that the contours do not surround any holes) must be zero. This argument explains in a very straightforward way one of the early paradoxes of superconductivity, that of flux exclusion, or the Meissner effect.

A.3 Gauge Transformations in Superconductivity

In some areas of electromagnetism it is convenient to derive the electric and magnetic fields of Maxwell's equations in terms of an electric scalar potential V and a magnetic vector potential A. As we discussed in the previous section, in all classical cases it is the fields which completely describe the physical features of any system, and these fields are unchanged if the potentials undergo the following transformations:

$$\begin{aligned} V' &= V - \mathrm{d}\chi/\mathrm{d}t \\ A' &= A + \nabla\chi \end{aligned} \tag{A.13}$$

where χ is any scalar quantity. This property of the potentials, known as gauge invariance, can be used as the starting point for deriving the form of the canonical momentum, using the requirement that the Hamiltonian should be invariant under such a 'gauge transformation'. The transformation to a quantum mechanical description leaves us with a gauge invariant Hamiltonian operator. However the solutions to Schrödinger's equation are modified by the change of gauge so that the new solution is related to the former by

$$\psi' = \psi \exp(-\mathrm{i}e\chi/\hbar). \tag{A.14}$$

A.3.1 Gauge symmetry breaking

At the basis of any symmetry principle in physics is the requirement that some physical quantity is not measurable (for example the absolute position in space is not measurable for a system which is invariant under spatial transformation). Most phase transitions involve some symmetry breaking by the ordered phase (for example the ferromagnetic transition involves the breaking of rotational symmetry). In the case of superconductivity the previously unmeasurable quantity is the local phase of the order parameter and the particular symmetry which is broken below T_c is local gauge symmetry. The free energy of the system is independent of the phase of the order parameter so that this quantity, as well as the Hamiltonian, is gauge invariant. On the other hand, a state with a well defined phase, rather than an ensemble of states with all possible values of phase, is not gauge invariant. If gauge invariance is applied to the superconducting state then the phase could be locally chosen arbitrarily by adding a scalar quantity $\chi(r)$ which can vary spatially. The essential feature of the Ginzburg–Landau equation is that an energy cost is introduced by a spatial gradient of the phase of the order parameter (see equation (1.10)). Thus local gauge invariance no longer applies to the superconducting state. (Global gauge invariance, by which a universal constant phase factor can be added to all wave functions, is still a good symmetry, even in superconductors.)

A.3.2 Other mechanisms for modifying the phase

In addition to the scalar or vector potentials, the phase of the superconducting order parameter can also be modified by physical rotation. Reverting to the path integral formulation of quantum mechanics, the basic principle says that any additional contribution to the action of the system δS will multiply the amplitude in going from a point a to point b by the factor

$$(1/\hbar) \int \boldsymbol{p}' \cdot \mathrm{d}\boldsymbol{l} \tag{A.15}$$

where $\boldsymbol{p}'$ is the additional mechanical momentum arising from the rotation. The total angular momentum about the axis of rotation is $\boldsymbol{L} = \int \boldsymbol{r} \times \boldsymbol{u}'(r)\rho \, \mathrm{d}V$ where ρ is the mass density of the body, $\boldsymbol{u}'(r)$ is the velocity field and the integral is over the volume of the body. When a circular superconducting ring is rotated with angular velocity ω, the additional rotational contribution to the phase shift in going from the point a back to point a becomes

$$\delta\theta = 2m\omega S/\hbar$$

where S is the cross-sectional area of the ring perpendicular to the axis of rotation and $2m$ is the mass of each electron pair. In order to satisfy the single-valued constraint on the order parameter for a contour of integration going once around the ring this contribution must be added to equation (1.15) so that the flux quantisation condition is modified to become

$$\Phi + 2m\omega S/e = n\Phi_0. \tag{A.16}$$

Thus for a rotating superconducting ring magnetic flux is not quantised. The rotation generates an internal additional magnetic field $\boldsymbol{B} = 2m\omega/e$ which is known as the London moment. One of the most interesting features of the London moment is that it is accurately aligned with the axis of rotation. The phenomenon has been applied to a measurement of the Compton wavelength of the electron (section 5.10), and to two different types of superconducting gyro (see section 8.1.1).

A.3.2.1 Gravitational phase shifts

Apart from rotational fields, applied gravitational fields give rise to phase shifts around a superconducting ring so that, at least in principle, a gravity gradiometer may be built (Brady 1983).

Appendix B

Mechanical Analogue of the Classical Josephson Effects

B.1 The Tilted Washboard Analogue of the Josephson Effects

The 'conventional' Josephson effects of Chapter 2 can be most succinctly described in terms of a single second-order non-linear differential equation which describes the time evolution of the order parameter phase difference φ across the Josephson junction. For a junction with dimensions small compared with λ_J and critical current i_1, shunted by an ohmic resistance R and a capacitance C, the equation of motion can be written

$$\frac{\hbar C}{2e}\frac{d^2\varphi}{dt^2}+\frac{\hbar}{2eR}\frac{d\varphi}{dt}+i_1\sin\varphi=i. \tag{B.1}$$

Equation (B.1) can only be solved numerically, except in one or two particular situations, specifically when there is no voltage across the junction, so that φ does not increase steadily with time but only oscillates about some mean value. The numerical solutions are quite straightforward and figures 4.4–4.6 show some typical examples of the way in which the phase and voltage evolve with time for the more complex system with two junctions in parallel. In this section we are more concerned with trying to give some physical insight into the behaviour of Josephson junctions and SQUIDs by treating a mechanical analogue of the systems. This is based on the so-called 'tilted washboard' model, introduced by Anderson (1964).

Imagine a ball bearing sitting in a groove on a washboard or other corrugated surface (see figure B.1). If the washboard is tilted along a line running at right angles to the corrugations, initially for small angles of tilt the ball bearing will remain in the same corrugation, merely adjusting its position near the bottom of the depression. However once some critical slope angle is reached the ball bearing will move into the next depression and then continue rolling down the washboard, at an accelerating rate. The analogue of the distance which the ball bearing has moved along the washboard is the phase difference across the junction. The critical slope corresponds to the critical current of the junction and the angle of inclination is the analogue of the current being driven from an external source through the junction. Thus for low currents (angles) the phase across the junction (position of the ball bearing) is independent of time, except perhaps that the phase can oscillate about its mean value (the ball can oscillate around in the bottom of a hollow). When the current exceeds the critical current (critical tilt angle) the phase difference begins to increase with time (the ball begins to roll). In real Josephson junctions the phase difference rapidly reaches a constant rate of increase (ball reaching its terminal velocity, due to friction between it and the washboard). Once there is a finite voltage across the junction (i.e. the ball rolling, since the ball velocity is the analogue of the direct junction voltage), then a reduction of the current below the critical value may not cause the phase difference to stop increasing with time, due to the capacitance of the junction (inertia of the ball).

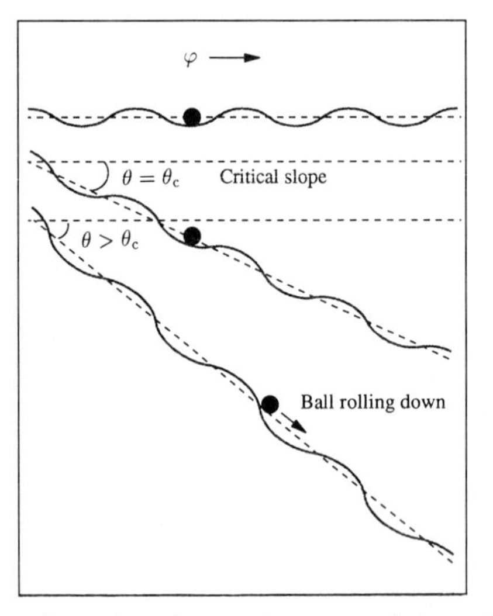

Figure B.1 Josephson junction analogue consisting of a ball rolling on a corrugated inclined plane

It is straightforward to write down the internal energy for the two systems of junction and ball bearing, to make direct analogies between terms in each. For the ball bearing there is first the potential energy due to the vertical height of the ball and the acceleration, g, due to gravity:

$$U_1 = -mgz = -mgh \cos \varphi$$

where $2h$ is the peak-to-trough washboard height. The corresponding term for the junction is

$$U_1 = -(i_1\Phi_0/2\pi)\cos\varphi$$

and this represents the coupling energy of the order parameter across the junction (see section 2.1) where here i_1 is the critical current of the junction. For a junction biased by an external current i_0 there is an additional contribution to the internal energy of amount

$$U_0 = i_0\Phi_0\varphi.$$

The physical origin of this term is simply derived. Suppose the current is great enough to bias the junction into the finite voltage regime. Then the junction power dissipation is

$$P(t) = Vi_0 = i_0\Phi_0\,\mathrm{d}\varphi/\mathrm{d}t.$$

Integrating this expression with respect to time to find the corresponding internal energy term in the general situation when the phase may or may not be time-dependent gives

$$U_0 = \int P\,\mathrm{d}t = \Phi_0\varphi i_0$$

where, as usual with potential energy terms, the origin is arbitrary. Thus the total internal energy expression for a single isolated junction becomes

$$U = U_1 + U_0 = -\Phi_0 i_1[\cos\varphi + (i_0/i_1)\varphi]. \tag{B.2}$$

In the analogous case of the ball bearing it is clear that a term in the internal energy $U = mg\varphi\tan\theta$ exists, where θ is the angle of inclination of the washboard to the horizontal. Thus the washboard slope is the analogue of the bias current. (Actually to get exact agreement it is necessary to stretch (literally) the washboard, to maintain its 'wavelength' along the horizontal axis as it is tilted.)

Consider the tilt angle increasing steadily from zero. The ball bearing will just begin to roll when, for all points on the path, $\mathrm{d}U/\mathrm{d}\varphi > 0$. The required condition is that $\tan\theta > h\sin\varphi$ for all values of φ, i.e. $\tan\theta > h$. Similarly for the Josephson junction the phase difference will begin to increase continuously with time when $\mathrm{d}U/\mathrm{d}\varphi > 0$ everywhere, corresponding to

$$i_0 > \max(i_1\cos\varphi) = i_1$$

which, of course, is the condition that we have already used as a definition of the critical current of the junction. To prevent the ball bearing, once rolling, from accelerating without limit it is necessary to introduce some friction between it and the corrugated surface, so that there is a force term acting on it, in the opposite sense to its direction of motion and of magnitude proportional to the ball's velocity. The work done against this force per unit time is just the power dissipated by the ball and is $F\,\mathrm{d}\varphi/\mathrm{d}t = kv^2$ where v is the mean velocity and k is the dissipation constant. Similarly the junction voltage V is the analogue of the phase velocity $\mathrm{d}\varphi/\mathrm{d}t$ and the power dissipated is V^2/R where R is the phenomenological resistance of the junction when it is biased into the finite voltage state.

Finally we see that once the ball is moving it possesses kinetic energy which may be enough to allow it to continue to surmount local potential energy maxima, which it would not be able to do if initially stationary. Thus tilting the washboard and then reducing the slope to less than its critical value may allow the ball to continue rolling. Analogously the kinetic energy of the junction is $CV^2/2$, where C is the junction capacitance. The condition for such 'hysteretic' behaviour in the case of the ball bearing is clearly that its kinetic energy at the bottom of each corrugation should be greater than the sum of the potential energy associated with the height of the corrugation and the energy lost through friction. Assuming that the ball rolls at a mean terminal velocity v, this condition can be written:

$$mv^2/2 > mgh.$$

But the terminal velocity v is related to the dissipation coefficient k through $v = mg\theta$ and combining these two results in the condition for hysteresis gives

$$m^2 g\theta^2/2k^2h > 1. \tag{B.3}$$

By analogy we can predict that the Josephson junction will show hysteretic behaviour if

$$2\pi i_1 R^2 C/\Phi_0 > 1 \tag{B.4}$$

and this is borne out experimentally. Hysteretic behaviour in junctions manifests itself through a double-valued current–voltage characteristic, as shown in figure 2.3(*b*). Thus the critical current for an increasing bias current is greater than that observed when the bias current is being reduced.

B.2 Washboard Analogue of a DC SQUID

The tilted washboard analogy can be carried over to SQUIDs as well. In this case it is also necessary to consider the internal energy, U, associated with the circulating supercurrent, i, in the ring, $U = Li^2/2$. With a single-junction SQUID it is not possible to establish a direct voltage across the junction by applying a direct current. However it is possible to apply a linear flux ramp which has the same effect. The additional circulating current energy term turns the tilted washboard into a corrugated parabola. Application of an alternating flux corresponds to rocking this parabola backwards and forwards, so that the lowest-energy state changes from corrugation to corrugation, in step with the external flux. Figure 3.1 shows these energy surfaces for an RF SQUID for two different values of applied DC flux.

It is instructive to try to apply our washboard analogue to the case of the DC SQUID. In this case we have two junctions biased so that each has the same time-averaged voltage, or analogously, the same average washboard slope. The relative phases of the adjacent washboards, representing the phase changes across each junction, are adjusted by relative displacement of the two, and this corresponds to a change in the flux applied to the SQUID ring.

The junctions are not of course uncoupled. The total phase change around the ring (see section 4.2, equations (4.9)–(4.12)), including that due to the flux in the ring as well as the phase changes across the junctions, must sum to $2n\pi$. Thus if the phase change across one junction changed independently of the other there would have to be a change in circulating current. The total internal energy of the SQUID now contains the following terms: (i) the Josephson coupling energies of the junctions, (ii) the energy supplied by the external bias circuit and (iii) the magnetic energy of the ring associated with the circulating supercurrent. The relative phase changes, together with the circulating current, adjust themselves to minimise this total internal energy. In the washboard analogue this could be modelled by joining the ball bearings, rolling on their separate washboards, by a light Hooke's law spring, to represent the inductive energy of the ring. Equation (2.8) dictates that a change in circulating current δi would produce a phase change $\delta\varphi$ in the order parameter. The energy change in this case is $\delta U = L\delta i^2/2 \propto \delta\varphi^2$, or in the case of the analogue, the square of the distance separating the balls, along the horizontal. This is of course the way in which the stored energy in a spring changes with extension.

When the balls are rolling down their respective slopes, their horizontal separation oscillates, depending on the relative position of each ball with respect to its local trough, at the Josephson frequency. The mean separation depends on applied flux. It is intuitively reasonable that the balls will always remain within a phase difference of 2π of one another, where this is the phase difference arising from the circulating current, rather than from the applied flux. This result is borne out by numerical calculations based on equations (4.9)–(4.12), which show that, even if a large phase difference is initially introduced, it rapidly dies away on a time scale of order R/L to settle down to a maximum phase variation less than 2π. This result is to be contrasted with the magnitude of the phase difference which can be introduced in the zero-voltage regime, which is of order Li_1/Φ_0, and can be much greater than 2π. However once the phase changes across the junctions become time-dependent, the dynamics of the system ensures that the circulating current oscillating at the Josephson frequency has a maximum amplitude $\pm\Phi_0/L$. This explains qualitatively the well known numerical result (Tesche and Clarke 1977) that the degree of modulation of the critical current of a DC SQUID is limited to this same magnitude, since the circulating supercurrent at the transition into the finite voltage regime can either add or subtract from the portion of bias current flowing through each junction and thus increase or decrease the effective critical current of the two junctions in parallel by about the same amount. Thus in some sense it may be said that, no matter what the apparent value of $\beta = 2\pi L i_1/\Phi_0$ for the DC SQUID, once biased into the operating regime it will behave as if its β value is close to unity, the optimum value.

References

Aitken M J 1972 *Physics and Archaeology* (Oxford: Oxford University Press)
Ambegaokar V and Baratoff A 1963 *Phys. Rev. Lett.* **10** 486–9
Anderson A C, Cabrera B, Everitt C W F, Leslie B C and Lipa J A 1982 *Proc. II Marcel Grossman Meeting on General Relativity* (Amsterdam: North Holland) pp 939–57
Anderson P W 1964 *Lectures on the Many Body Problem* vol 2 ed Caianello (New York: Academic)
Ashcroft D N and Mermin D W 1976 *Solid State Physics* (New York: Holt, Rinehart and Winston)
Awschalom D D, Rozen J R, Ketchen M B, Gallagher W J, Kleinsasser A W, Sandstrom R L and Bumble B 1989 *Appl. Phys. Lett.* **53** 2108–10
Bardeen J, Cooper L N and Schrieffer J R 1957 *Phys. Rev.* **108** 1175–204
Barone A and Paterno G 1981 *Physics and Applications of the Josephson Effect* (New York: Wiley–Interscience)
Bednorz J G and Müller K A 1986 *Z. Phys.* B **64** 189–93
Bermon S, Chaudhari P, Chi C C, Tesche C and Tsuei C C 1985 *Phys. Rev. Lett.* **55** 1850–3
Blair D G 1987 *Contemp. Phys.* **28** 457–75
Blaney T G 1978 *Future Trends in Superconductive Electronics* (*AIP Conf. Proc.*) ed B S Deaver p 44230
Blaney T G and Knight D J E 1974 *J. Phys. D: Appl. Phys.* **7** 1882–6
Bloch F 1970 *Phys. Rev.* B **2** 109–21
Brady R M 1983 *Thesis* Cambridge University (unpublished)
Cabrera B 1982 *Phys. Rev. Lett.* **48** 1378–81
Caldeira A O and Leggett A J 1981 *Phys. Rev. Lett.* **46** 211–4
Campbell W H and Zimmerman J E 1975 *Geophysics* **40** 269–84
Carelli P and Foglietti V 1982 *J. Appl. Phys.* **53** 7592–8
Casimir H B G 1940 *Physica* **7** 887
Cerdonio M, Wang R H, Rossman J R and Mercereau J E 1973 *Proc. LT11 (Boulder, Colorado)* pp 525–32
Chambers R G 1960 *Phys. Rev. Lett.* **5** 3–5
Chiao R Y 1979 *IEEE Trans. Mag.* **MAG-15** 446–9
Claassen J H 1975 *J. Appl. Phys.* **43** 2268–75

Clark T D 1968 *Phys. Lett.* **27A** 585–6
Clark T D 1987 *Quantum Implications* ed B J Hiley and F D Peat (London and New York: Routledge and Kegan Paul)
Clarke J 1966 *Phil. Mag.* **13** 115–27
—— 1972 *Phys. Rev. Lett.* **28** 1363–9
—— 1983 *IEEE Trans. Mag.* **MAG-19** 288–94
Clarke J, Goubau W M and Ketchen M B 1976 *J. Low Temp. Phys.* **25** 99–144
Clarke J and Hawkins G 1975 *IEEE Trans. Mag.* **MAG-11** 84–9
Cohen D, Edelsack E A and Zimmerman J E 1970 *Appl. Phys. Lett.* **16** 278–80
Colclough M S, Gough C E, Keene M, Muirhead C M, Thomas N, Abell J S and Sutton S 1987 *Nature* **328** 47–9
Cooper L N 1956 *Phys. Rev.* **104** 1189
Cromar M W and Carelli P 1981 *Appl. Phys. Lett.* **38** 723–5
Day E P 1972 *Appl. Phys. Lett.* **29** 540–2
Deaver B S and Fairbank W M 1961 *Phys. Rev. Lett.* **7** 43–6
Dekel A and Rees M J 1987 *Nature* **326** 455–62
de Waele A Th A M and de Bruyn Ouboter R 1969 *Physica* **42** 225–54
Donaldson G B and Bain R J 1984 *Appl. Phys. Lett.* **45** 990–2
Drever R W 1961 *Phil. Mag.* **6** 683–7
Duret D and Karp P 1983 *Nuovo Cimento* D **2** 123–41
Edelsack E A 1973 *The Science and Technology of Superconductivity* ed W N Gregory, W N Mathews and E A Edelsack (New York: Plenum)
Ehnholm 1977 *J. Low Temp. Phys.* **29** 1–27
Elsley R K and Sievers A J 1974 *Rev. Phys. Appl.* **9** 295
Fagaly R L 1989 *Supercond. Ind.* **2** 24–30
Fairbank W M 1977 *Weak Interaction Physics* (*AIP Conf. Publ. 37*) p 51
Ferrari M J, Johnson M, Wellstood F, Clarke J, Inam A, Wu X D, Nazar L and Venkatesan T 1989 *Nature* **341** 723–5
Finnegan T F and Wahlsten S 1972 *Appl. Phys. Lett.* **21** 541–4
Fiory A T 1971 *Phys. Rev. Lett.* **27** 501–3
Foglietti V, Koch R H, Gallagher W J, Oh B, Bumble B and Lee W Y 1989 *Appl. Phys. Lett.* **54** 2259–61
Frederick N V, Sullivan D B and Adair R T 1977 *IEEE Trans. Mag.* **MAG-13** 361–4
Frisch H 1990 *Nature* **344** 706–7
Frolich H 1950 *Phys. Rev.* **79** 845
Fulton T A 1973 *Phys. Rev.* B **7** 981–5
Fulton T A and Dolan G J 1987 *Phys. Rev. Lett.* **59** 109–12
Gallop J C 1973 *NPL Rep.* QU 22
—— 1976 *J. Phys. D: Appl. Phys.* **9** 2111–5
—— 1977 *Proc. Int. Conf. on SQUIDs* (*Berlin 1976*) pp 267–71
Gallop J C, Langham C D, Radcliffe W J and Roys W B 1988 *Phys. Lett.* **128A** 222–4
Gallop J C and Petley B W 1974 *IEEE Trans. Instrum. Meas.* **IM-23** 267–70
—— 1976 *J. Phys. E: Sci. Instrum.* **9** 417–29
—— 1983 *Nature* **303** 53–4
Gallop J C and Radcliffe W J 1981 *Physica* B **107** 621–2
Gallop J C, Radcliffe W J, Langham C D, Sobolewski R, Kula W and Gierlowski P 1989 *Supercond. Sci. Technol.* **2** 1–4

Geerligs L J, Anderegg V F, Holweg P A M, Mooij J E, Pothier H, Esteve D, Urbina C and Devoret M H 1990 *Phys. Rev. Lett.* **64**
Geerligs L J, Peters M, de Groot L E M, Verbruggen A and Mooij J E 1989 *Phys. Rev. Lett.* **63** 326–9
Giaver I and Zeller H R 1968 *Phys. Rev. Lett.* **20** 1504–7
Giffard R P, Gallop J C and Petley B W 1977 *Prog. Quantum Electron.* **4** 301–402
Giffard R P, Webb R A and Wheatley J C 1972 *J. Low Temp. Phys.* **6** 533–611
Ginzburg V L and Landau L D 1950 *Zh. Eksp. Teor. Fiz.* **20** 1064
Goldstein H 1969 *Classical Mechanics* (Reading, MA: Addison-Wesley)
Goree W 1972 *Proc. Appl. Supercond. Conf.* IEEE Publ. 72CH0682 pp 640–8
Goree W S and Fuller M 1976 *Rev. Geophys. Space Phys.* **14** 591–608
Gor'kov L P 1959 *Sov. Phys.—JETP* **36** 1364–7
Gough C E, Colclough M S, Forgan E M, Jordan R G, Keene M, Muirhead C M, Rae A I M, Thomas N, Abell J S and Sutton S 1987 *Nature* **326** 855
Grimes C C and Shapiro S 1968 *Phys. Rev.* **169** 397–406
Grimes D, Lennard R and Swithenby S 1983 *Nuovo Cimento* D **2** 650–9
Gross R, Chaudhari P, Kawasaki M, Ketchen M B and Gupta A 1990 *Appl. Phys. Lett.* **57** 727–9
Gutmann P and Bachmair H 1989 *Superconducting Quantum Electronics* ed V Kose (Berlin: Springer)
Hamilton C A 1972 *Phys. Rev.* B **5** 912–23
Hamilton C A, Lloyd F L and Kautz R L 1985 *IEEE Trans. Mag.* **MAG-21** 197–9
Hamilton C A, McDonald D G, Sauvageau J E and Whiteley S R 1990 *Proc. IEEE* **77** 1224–32
Hartland A 1981 *Precis. Meas. Fundam. Constants* **2** (*NBS Spec. Publ. 617*) 543–8
Hartland A, Finnegan T F, Witt T J and Reymann D 1978 *IEEE Trans. Instrum. Meas.* **IM-27** 470–4
Hartwig H W and Passow C 1972 *Applied Superconductivity* vol 2 ed V L Newhouse (New York: Academic)
Harvey I 1972 *Rev. Sci. Instrum.* **43** 1626–9
Hasuo S and Imamura T 1989 *Proc. IEEE* **77** 1177–93
Hilbert C and Clarke J 1983 *Appl. Phys. Lett.* **43** 694–6
Hirschkoff E C, Symko O G, Vant-Hull L L and Wheatley J C 1970 *J. Low Temp. Phys.* **2** 653–65
Hollenhorst J N and Giffard R P 1979 *IEEE Trans. Mag.* **MAG-15** 474–7
Hu Qing and Richards P L 1989 *Appl. Phys. Lett.* **55** 2444–6
Huang H C W and Schad R G 1983 *J. Appl. Phys.* **54** 3878–85
Hughes V W, Robinson H G and Beltran-Lopez V 1960 *Phys. Rev. Lett.* **4** 342–4
Inam A, Hegde M S, Wu X D, Venkatesan T, England P, Miceli E W, Chase C C, Chang J M, Tarascone J M and Wachtman J B 1988 *Appl. Phys. Lett.* **53** 908–10
Irie A, Era M, Yamashita T, Kurosawa H, Yamane H and Hirai T 1989 *Japan. J. Appl. Phys.* **28** L1816–9
Jackel L D, Kurkijarvi J, Luken J E and Webb W W 1972 *Proc. LT13* **3** 705–8
Jackel L D, Webb W W, Lukens J E and Pei S S 1974 *Phys. Rev.* B **9** 115–8
Jacklevic R C, Lambe J S, Silver A H and Mercereau J E 1964 *Phys. Rev. Lett.* **12** 274–5
Jain A K, Lukens J E and Tsai J S 1987 *Phys. Rev. Lett.* **58** 1165–8

Jaycox J M and Ketchen M B 1981 *IEEE Trans. Mag.* **MAG-17** 400–3
Josephson B D 1962 *Phys. Lett.* **1** 201–3
—— 1964 *Rev. Mod. Phys.* **36** 216–20
Kamper R A and Zimmerman J E 1971 *J. Appl. Phys.* **42** 132–6
Kanter H and Vernon F L 1972 *J. Appl. Phys.* **43** 3174–83
Kautz R L, Hamilton C A and Lloyd F L 1987 *IEEE Trans. Mag.* **MAG-23** 883–900
Ketchen M B, Awschalom D D, Gallagher W J, Kleinsasser A W, Sandstrom R L, Rozen J R and Bumble B 1989 *IEEE Trans. Mag.* **MAG-25** 1212–5
Ketchen M B and Jaycox J M 1982 *Appl. Phys. Lett.* **40** 736–8
Kingston J J, Wellstood F C, Lerch P, Miklich A H and Clarke J 1990 *Appl. Phys. Lett.* **56** 189–91
Knuutila 1988 *EPS Workshop: SQUIDs, State of the Art, Perspectives and Applications (Rome)*
Koch R H, Gallagher W J, Bumble B and Lee W Y 1989 *Appl. Phys. Lett.* **54** 951–3
Koch R H, van Harlingen D J and Clarke J 1981 *Phys. Rev. Lett.* **47** 1216–9
——1981 *Appl. Phys. Lett.* **38** 380
Kosaka S, Nagakawa H, Kawamura H, Okada Y, Hamazaki Y, Aoyagi M, Kurosawa I, Shoji A and Takada S 1989 *IEEE Trans. Mag.* **MAG-25** 789–4
Kotani S, Fujikama N, Imamura T and Hasuo S 1989 *Digest of Technical Papers, Int. Conf. on Solid State Circuits* pp 150–1
Kurkijarvi J 1972 *J. Appl. Phys.* **44** 3729–33
Laibowitz R B, Koch R H, Gupta A, Koren G, Gallagher W J, Foglietti V, Oh B and Viggiano J M 1990 *Appl. Phys. Lett.* **56** 686–8
Landau L D and Lifshitz E M 1972 *Quantum Mechanics* (Oxford: Pergamon)
Leggett A J 1980 *Prog. Theor. Phys. Suppl.* **69** 80
Leggett A J and Garg A 1985 *Phys. Rev. Lett.* **54** 857
Legros M, Kotlicki A, Crooks M J C, Turrell B G, Drukier A K and Spergel D N 1988 *Nucl. Instrum. Methods* A **263** 229–32
Levinsen M T, Chiao R Y, Feldman M J and Tucker B A 1977 *Appl. Phys. Lett.* **31** 776–8
Lhota J R, Scheuermann M, Kuo P K and Chen J T 1984 *Appl. Phys. Lett.* **44** 255–7
Likharev K K 1979 *Rev. Mod. Phys.* **51** 101–59
Likharev K K and Zorin A B 1985 *J. Low Temp. Phys.* **17** 347–60
Lindelof P E 1981 *Rep. Prog. Phys.* **44** 949–1026
Little W A 1987 *Proc. Symp. on Low Temperature Electronics and HTS* (Pennington, NJ: Electrochemical Society) pp 598–602
London F 1950 *Superfluids* vol 1 (New York: Wiley)
Long A P, Clark T D and Prance R J 1980 *Rev. Sci. Instrum.* **51** 8–13
Lumley J M, Somekh R E, Evetts J E and James J H 1985 *IEEE Trans. Mag.* **MAG-21** 539–42
Macfarlane J 1973 *Appl. Phys. Lett.* **22** 549–42
Mager A 1982 *Naturwissenschaften* **69** 383–8
Manley J M and Rowe H E 1956 *Proc. IRE* **44** 904
Martinis J M, Devoret M H and Clarke J 1987 *Phys. Rev.* B **35** 4682–98
Matisoo J 1967 *Proc. IEEE* **55** 172–80

McCumber D E 1968 *J. Appl. Phys.* **39** 3113–8
McDonald D G 1987 *Appl. Phys. Lett.* **50** 775–7
Meredith D J, Pickett G R and Symko O G 1973 *Phys. Lett.* **42A** 13–4
Muck M and Heiden C 1989 *IEEE Trans. Mag.* **MAG-25** 1151–3
Nagatsuma T, Enpuku K, Irie F and Yoshida K 1983 *J. Appl. Phys.* **54** 3302–9
Niemeyer J, Grimm L, Meier W, Hinken J H and Vollmer E 1985 *Appl. Phys. Lett.* **47** 1222–3
Ohta H, Takahata M, Takahashi Y, Shinada K, Yamada Y, Hanasaka T, Uchikawa Y, Kotani M, Matsui T and Komiyama B 1989 *IEEE Trans. Mag.* **MAG-25** 1018–21
Onnes K 1911 *Akad. van Wetenschappen* **14** 113
Ono R H, Cromar M W, Kautz R L, Soulen R J, Colwell J H and Fogle W E 1987 *IEEE Trans. Mag.* **MAG-23** 1670–3
Paik H J 1980 *Nuovo Cimento* B **55** 15–36
Parker W H, Langenberg D N, Denenstein A and Taylor B N 1969 *Phys. Rev.* **177** 639–64
Pedersen N F, Finnegan T F and Langenberg D N 1972 *Phys. Rev.* B **6** 4151–9
Petley B W 1989 private communication
Petley B W, Morris K A, Clarke R N and Yell R W 1976 *Electron. Lett.* **12** 237–8
Philips J D, Fairbank W M and Navarro J 1988 *Nucl. Instrum. Methods* A **264** 125–30
Pierce J M, Opfer J E and Rorden L H 1974 *IEEE Trans. Mag.* **MAG-10** 599–602
Prance R J, Clark T D, Long A P and Moore M 1981 *Cryogenics* **21** 501–6
Prance R J, Clark T D, Mutton J E, Prance H, Spiller T P and Nest R 1985 *Phys. Lett.* **107A** 133
Quate C F 1986 *Phys. Today* August
Richards M G, Cowan B P, Secca M F and Machin K 1988 *J. Phys. B: At. Mol. Phys.* **21** 665–81
Richards P L and Sterling S A 1969 *Appl. Phys. Lett.* **14** 394–6
Rieger T J, Scalapino D J and Mercereau J E 1971 *Phys. Rev. Lett.* **27** 1787–90
Robinson F N H 1972 *Noise and Fluctuations in Electronic Circuits* (Oxford: Oxford University Press)
Rogers C T and Buhrman R A 1984 *Phys. Rev. Lett.* **53** 1272–5
Schouten J S, Caplin A D, Guy C N, Hardiman M, Koratzinos M and Steer W S 1987 *J. Phys. E: Sci. Instrum.* **20** 850–61
Shapiro S 1963 *Phys. Rev. Lett.* **11** 80–2
Shiota T, Takechi K, Takai Y and Hayakawa H 1989 *Advances in Superconductivity* ed K Kitazawa and T Ishiguro (Tokyo: Springer) pp 755–9
Silver A H, Jaklevic R C and Lambe J 1966 *Phys. Rev.* **141** 367
Silver A H and Sandell R D 1983 *IEEE Trans. Mag.* **MAG-19** 623–5
Silver A H and Zimmerman J E 1967 *Phys. Rev.* **157** 317–41
Skocpol W J, Beasley M R and Tinkham M 1974 *J. Low Temp. Phys.* **16** 145–67
Sleator T, Hahn E L, Hilbert C and Clarke J 1985 *Phys. Rev. Lett.* **55** 1742–5
Soulen R J and Marshak H 1972 *Proc. Conf. on Applied Superconductivity* IEEE Publ. 72CH0682 pp 588–91
Stevens R J 1984 *Thesis* Lancaster University unpublished
Stewart W C 1968 *Appl. Phys. Lett.* **12** 277–80
Taber R 1974 *Thesis* Stanford University (unpublished)

Tassie L J 1965 *Nuovo Cimento* **38** 1935–7
Taylor B N, Parker W H and Langenberg D N 1969 *Rev. Mod. Phys.* **41** 375–440
Tesche C D 1990 *Phys. Rev. Lett.* **64** 2358–61
Tesche C D and Clarke J 1977 *J. Low Temp. Phys.* **29** 301–31
Tinkham M 1975 *Introduction to Superconductivity* (New York: McGraw-Hill)
Triscone J M, Fischer O, Brunner O, Antognazza L, Kent A D and Karkut M G 1990 *Phys. Rev. Lett.* **64** 804–7
Tsai J S, Jain A K and Lukens J E 1983 *Phys. Rev. Lett.* **51** 316–9
Tucker J R and Millea J F 1979 *IEEE Trans. Mag.* **MAG-15** 288–90
Turneaure J E, Everitt C W F, Parkinson B W, Bardas D, Breakwell J V, Buchman S, Cheung W S, Davidson D E, DeBra D B, Fairbank W M, Feteih S, Gill D, Hacker R, Keiser G M, Lockhart J M, Muhlfelder B, Parmley R T, Xinhua Y M and Zhou P 1989 *Adv. Space Res.* **9** 29–38
Van Duzer E and Turner C W 1981 *Principles of Superconducting Electronics* (London: Edward Arnold)
Venkatesan T, Inam A, Dutta B, Ramesh R, Hegde M S, Wu X D, Nazar L, Chang C C, Barner J B, Hwang D M and Rogers C T 1990 *Appl. Phys. Lett.* **56** 391–3
Venkatesan T, Wu X, Inam A, Chang C C, Hegde M S and Dutta B 1989 *IEEE J. Quantum Electron.* **25** 2388–93
Voss R F, Laibowitz R B, Raider S I and Clarke J 1980 *J. Appl. Phys.* **51** 2306–9
Voss R F and Webb R A 1981 *Phys. Rev.* B **24** 7447–9
Wada Y 1990 *Proc. IEEE* **77** 1194–207
Wahlsten S, Rudner S and Claeson T 1978 *J. Appl. Phys.* **49** 4248–63
Webb R A, Giffard R P and Wheatley J C 1973 *J. Low Temp. Phys.* **13** 383–405
Wellstood F C, Kingston J J and Clarke J 1990 *Appl. Phys. Lett.* **56** 2336–8
Wellstood F C, Urbina C and Clarke J 1989 *Appl. Phys. Lett.* **54** 2599–602
Wheatley J C 1975 *Rev. Mod. Phys.* **17** 415–70
Whiteley S R, Hohenwarter G K G and Faris S M 1986 *IEEE Trans. Mag.* **MAG-23** 899–902
Widom A 1979 *J. Low Temp. Phys.* **37** 449
Will C M 1981 *Theory and Experiment in Gravitation Physics* (Cambridge: Cambridge University Press)
Witt T J 1988 *Phys. Rev. Lett.* **61** 1423–6
Wolf P, van Zeghbroek B J and Deutsch U 1985 *IEEE Trans. Mag.* **MAG-21** 226–9
Wu M K, Ashburn J R, Torng C J, Hor P H, Meng R L, Gao L, Huang Z J, Wang Y Q and Chu C W 1987 *Phys. Rev. Lett.* **58** 908–10
Wu X D, Xi X X, Li Q, Inam A, Dutta B, DiDomenico L, Weiss C, Martinez J A, Wilkens B J, Schwarz S A, Barner J B, Chang C C, Nazar L and Venkatesan T 1990 *Appl. Phys. Lett.* **56** 400–2
Zimmerman J E 1967 *Appl. Phys. Lett.* **10** 193–5
—— 1972 *J. Appl. Phys.* **42** 4483–7
—— 1977 *J. Appl. Phys.* **48** 702–10
—— 1980 *SQUID 80* (Berlin: Walter de Gruyter) pp 423–43
Zimmerman J E, Thiene P and Harding J T 1970 *J. Appl. Phys.* **41** 1572–80
Zimmerman J E, Beall J A, Cromar M W and Ono R H 1987 *Appl. Phys. Lett.*

Index